GREENSPEAK

GREENSPEAK

Fifty Years of Environmental Muckraking and Advocacy

Michael Frome

THE UNIVERSITY OF TENNESSEE PRESS / Knoxville

This book is printed on recycled, acid-free paper.

Library of Congress Cataloging-in-Publication Data

Frome, Michael.
Greenspeak: fifty years of environmental muckraking
    and advocacy/Michael Frome.—1st ed.
      p. cm.
ISBN 1-57233-177-1 (pbk.: alk. paper)
1. Environmentalism.
2. Conservation of natural resources.
3. Environmental protection.
I. Title.
GE195 .F76 2002
363.7'0525—dc21   2002004365

For Stewart Brandborg, Brock Evans, and Mack
Prichard, my close friends and allies,

> *Compañeros de mi vide, para*
> *guerrida de aquellos tiempos.*

But I won't say *Adios, muchachos,* because
they are going strong, still showing the way.

Contents

Illustrations

Preface

Notes written on motel stationery and on the backs of envelopes meant for use in public speaking kept turning up in desk drawers and inside of books on shelves, and so did copies of speeches filed in cabinets and inside my computer, reminding me of an abundance of words spoken over a period of forty-plus years. In a dream one night they all paraded across thin air before me, then marched to the ceiling overhead, where they fell apart and cascaded in thousands of separate sheets over the bed and floor. They lay still, but I felt mandated to recover and restore them, which I began to do. Suddenly I awoke, sat straight up in bed, and asked aloud, "But who cares?" I turned over and fell asleep, but in the morning I asked the question again and instinctively answered, "Well, I care." Yes, yes, I do care enough to put this collection together.

The first of the speeches to follow was given in 1963, the last in the year 2001. I talked in conference halls before citizen environmentalists and resource professionals, on college and university campuses, and maybe best of all in the out-of-doors. I'm now twice the age I was when I gave the first speech, and changed physically, a step slower and puffing on the upgrade, but maybe improved and matured in other ways, and the reader hopefully will find something constructive and consistent in what I've said. The speeches follow chronological order, except for the penultimate, which somehow refused to be regimented and then seemed to fit perfectly right where it is.

But then, just as I thought I had everything under control and organized, I discovered more materials in my files and more questions inside my head. I asked myself about circumstances surrounding the speeches and the people connected with them. In hindsight, what would I have said differently? How have I changed and evolved as the world changed and revolved around me? What have I learned, what do I have worthwhile to share? In a 1975 article about me *Time* magazine wrote, "Frome has made enemies in big business, the gun lobby and on Capitol Hill." Did I really go about making enemies? If true, did I have any friends and who were they? And did we accomplish anything of importance to record and remember?

While sorting all this out, I received a critique of my work, which I had requested, from Todd Wilkinson, a mature young (well, under forty) professional writer. One year earlier, in 1999, I had sat down with Todd at Rocky Mountain Roastings in Bozeman, Montana, where he lives. That cup of coffee stretched out over three hours, during which time we examined and compared each other's goals, illusions, disappointments, and disillusions. Along the way I mentioned the speeches I had given, and Todd urged me to assemble and publish them. Now, in reviewing the manuscript, he wrote,

> I think the speeches are excellent references from your many years as an activist, but what they need, in my mind, are either transitions or more cohesion; you know this far better than me, having spent time observing the way "young people" think in the classroom, but many in fact need to have their hands held as they are walked through the modern history of the conservation movement. They have passion, but they don't have a profound sense of what has come before them.
>
> Now more than ever, and especially with David Brower's passing [in November 2000], we need to be reminded that many of the sages remain with us who were in the trenches and possess the volume of knowledge, courage, and experience that carried the movement forward.
>
> What I would further enjoy reading is a chapter of short "Fromeian" character sketches that look back at the people you have known or who have brought inspiration to your pen (e.g., keyboard). I would like to hear you wax for a couple of pages on the giants of the past century, sharing thoughts on their vision as well as their

human qualities. We all love heroes, but more important, in my mind, is that we need to understand that
meaningful activism comes from people who are not
immortal and perfect but people of conviction who
took a stand for what they believe in, and that made
them remarkable.

I felt Todd's suggestions were cogent and helpful and so am including extrapolations, of varying length and scope, before and following the
speeches and a few interpolations within them for clarification or updating. I hope these will prove more helpful than obtrusive.

Another young writer friend, Dan Oko, in Austin, Texas, gave me further critique. He suggested that I organize the speeches along the lines
of issues rather than chronologically. Sections as he saw them would
include (1) Wilderness, (2) A Writer's Life, (3) Recreation and Land Use,
(4) Activism, and (5) Endangered Species and Other Animals. I considered Dan's idea; he certainly had the classifications right, but I decided
to stick with chronology, to show where and what I was doing at given
times in my life.

He raised another point:

I recall a couple of years ago speaking with Barry
Lopez, who had just come out with what he considered
to be his first true autobiographical book in a career
spanning decades. Lopez told me that he had finally
realized that people were interested in not just what he
had done, but what he had to say—and who better to
reveal that story that the writer himself? You have a bit
of the opposite problem. People want to know not just
what Michael Frome has said but what he's done.

I don't know that I have to spell out what I've done, when, basically,
I want to contribute to the record of history of my time. I do not want
the personal story to get in the way.

My experience in the classroom has shown me that history begins
for young people where they come in. I can illustrate this point in
another context. In 1998 I sent a copy of my then new book *Green Ink*
to Ben W. Gilbert, who had been city editor of the *Washington Post* when
I was a young reporter there following World War II. Now we were both
living on the West Coast, about two hundred miles apart in Washington

state, "the other Washington," and were corresponding via email. He was very kind when he wrote me: "The news business lost a very promising person when you left, but I think you are doing more good monitoring the environment." His reaction to *Green Ink* surprised me, considering that I spent considerable effort in debunking the mainstream media's shibboleth of objectivity. But no, he replied that implicit and explicit bias extended to more than coverage of the environment. He sent me a copy of a forty-four-page monograph he wrote in 1993, titled "A Recollection: Lifting the Veil from the 'Secret City'—The *Washington Post* and the Racial Revolution."

Gilbert's monograph sought to reconstruct the paper's institutional memory of its coverage of the racial revolution in the District of Columbia and the nation. He wrote:

> I reflected on how much each succeeding group of reporters and editors needs to know and appreciate about what went before to understand better what is happening and what may be promised in the crystal ball. . . .
>
> Newspapers mirrored pervasive segregation and entrenched racism. The *Washington Post* did not employ a single black reporter or editor nor could one of its reporters share a meal with a black person at downtown restaurants. That was academic because few staff members knew any members of the Capital's black community well enough for lunch.

I remember those days clearly. "Don't bother, it's black," the tough old night city editor would say, meaning we would not cover the event or happening, although crimes by blacks against whites were generously reported. And to identify black persons in the news by race was a traditional practice (which the *Washington Post* ultimately defied and demolished). But Washington was a southern city, controlled by segregationist southern politics in Congress. The YWCA was virtually the only place where whites and blacks ate in the same dining room. When a bus or streetcar crossed the Potomac into Virginia, the driver stopped and blacks moved to the rear. Such are the facts of history.

"Daily newspapers," wrote Gilbert, "were chameleons—as *white, black,* or *colorblind* as their readership, management, editors, reporters and prevailing social patterns suggested, which meant generally 'white.'"

He quoted in his monograph a recollection of Murrey Marder, who covered Washington racial issues in 1948 before transferring to national and international affairs: "For true-blue or ersatz Southerners, of whom we had our abundant share in editors as well as reporters, the notion that segregation was news was itself subversive. They were journalistically reared on the conviction that Negro news was not news except in mass-disaster figures or examples of the genetic stupidity or criminality of the race."

Having lived and worked in that period, I feel the sense of time, sense of place, and sense of what has gone on before, and I recognize the profound differences between then and now, all in my own working life. The *Washington Post* was far from the prestigious paper it became after the historic investigative stories about Watergate. It was published in a rickety, run-down, crowded building, competing for its market share with three other daily newspapers in the city. I began there in 1941 at the bottom, as a copy boy earning fifteen dollars per week. When I returned after the war as a reporter I earned thirty-five dollars per week. It was a marvelous learning experience. Journalism school graduates don't know what they're missing.

Years later, going around to speak here and there, I saw a lot of America and met many Americans who care about this country without dollar signs marked all over it. Once Eugene Odum, the pioneer ecologist, and I were on a program together, enabling me to listen and learn principles of biotic diversity. He was a scientist, not an activist, but he wanted me to speak to his graduate students at the University of Georgia, and later I did go there. Another time I was invited by Nancy Russell to speak in Portland, Oregon, to the annual meeting of the Friends of the Columbia River Gorge. She was a woman of stature in the community who enjoyed plants and flowers in general and the botanical wonderland of the Columbia River Gorge, which she was determined to save, so I benefited from a personally conducted tour. Then I benefited again when I was introduced at the Friends meeting by former senator Maurine Neuberger, who had succeeded her late husband, Richard Neuberger, both people who rose above the level of ordinary politicians.

I was introduced on a program in Washington by Rep. John Seiberling of Ohio, who was cut from the same cloth as the Neubergers. Seiberling said he was glad we could share the platform because I made him look like a moderate. I think that is what got me invited to some venues and thoroughly banned in others. In 1981, in a wilderness setting in the Grand Canyon, Martin Litton introduced me by saying, "Here's Mike

Frome, who was fired from every job he ever had," and everybody cheered. At Virginia Tech, at Blacksburg, Virginia, John Hosner, director of the Forestry and Wildlife Program, said with a smile that he was determined to have me come, even though it might cost the school a ten-thousand-dollar grant from the timber industry for new equipment. And maybe it did. On the appointed day, I spoke in two classes and then prepared to address the honors and awards banquet. Seated at the head table with me were an official of one of the big timber companies, Westvaco, and his wife. They were affable with chitchat—before I spoke. Later Hosner told me the lady complained that she had never been so insulted in her entire life as she had by my remarks.

I didn't mean to turn her off, or anyone else at any time, but I could never see the point of making a speech without something to say. Dick Neuberger recorded advice given to him by an older U.S. senator interested in his budding career: "Say nothing, and say it well." Neuberger did not play it safe. Neither did Sigurd Olson, William O. Douglas, Marjory Stoneman Douglas, and others—and neither, I felt, should I. When I went to speak at the honors and awards banquet at the West Virginia University forestry school, the alumni in the audience were first introduced—twenty or thirty, a whole crowd, sitting squarely in front of me, and almost all from the timber industry. It was intimidating, but I gave my best shot. When the banquet was over, a group of students took me for a beer and said, "They are here and around us all the time. We never get your viewpoint." When I spoke at Clemson in South Carolina, the president of the university sat at the head table. I felt some trepidation with him close at hand, but gave my speech anyway, including a roasting of the Army Corps of Engineers for the damage it did to rivers of the South and other regions. When I was through the president was the first to reach me. He shook my hand and thanked me, for he had been having trouble with the Corps on a project bordering the campus. It was a pleasant surprise, but that happens too.

For better or worse, I wrote my own speeches. People in power have someone else to do that work for them. Speech writing has become a profession in itself. It may not be honest, either for the writer or the speaker, but perhaps in politics or business it's simply the immediate effect that counts.

For example, when George Bush was running for president in 1988 he came to Seattle to deliver a speech on the environment to an environmentally conscious community. He declared as follows: "Let me say right at the outset that I don't think we've been doing enough to protect our environment in recent years. We need to do more. We are all passengers on a boat that we have damaged—not with the cataclysm of war, but with the slow neglect of a vessel that we thought was impervious to our abuse. In the last analysis, we all have a stake in maintaining the ecological health of the planet."

Bush said he was proud of Republican leadership in protecting the environment. He cited the record of Theodore Roosevelt, who eighty years earlier challenged Americans to face what Bush called "the great central task of leaving this land even a better land than it is for us."

Once elected, however, Bush did little if anything to face that central task and proved a poor environmental president. The speech in Seattle plainly was something served to him that he delivered and then forgot. It may be the way modern America is run and governed, but that doesn't make it valid.

Sometimes I was invited precisely because I was outspoken and critical. One day in 1991 L. W. (Bill) Lane Jr., whom I knew, called me on the telephone. He was the retired publisher of *Sunset Magazine,* a well-to-do Californian, generally conservative but truly supportive of the national parks. At the moment he was calling about the upcoming symposium at Vail, Colorado, marking the seventy-fifth anniversary of the National Park Service which he was helping to underwrite. Apparently registrations were low because people felt the speakers list lacked diversity. Lane invited me to come to inject a touch of criticism into what seemed like a safely stacked program. Personally, I felt that Vail, with its high-rise hotels and condominiums, and everything about it damaging the environment, was a poor and inappropriate setting for a conference on national parks, but I accepted the invitation. On reaching Vail, I learned that I was to be fourth on a panel of four moderated by Nathaniel P. Reed, a former assistant secretary of the interior, whom I knew very well and respected highly. I felt I was being used and protested to Reed; after all, the last speaker almost always has only five minutes, and by then people are heading for the exits. Reed insisted this was not the case, that he wanted me to be "the home-run hitter," and that I would have all the time I needed. He gave me a warm and affectionate introduction that surprised me and straightened my spine.

Of course, I remember the speech that I gave, but the lasting thrill was in the introduction, a reaffirmation that I had something worthwhile to say. That introduction and three others, spoken at different times and places, follow on the next pages.

Acknowledgments

Lisa Friend and Jim Spaich were the first to review the manuscript in the early stages and to provide helpful input. Both are former students of mine and gifted writers. Lisa additionally indexed the book with patience and proficiency (as she has with three of my earlier books).

Dan Oko and Todd Wilkinson, professional writers and friends, whose comments are cited in the preface, were the next reviewers. Then James R. Fazio, professor at the University of Idaho, and Ted Williams, editor-at-large of *Audubon* magazine, gave further input. Norma Grier, executive director of the Northwest Coalition for Alternatives to Pesticides, reviewed a section of the book dealing with chemical poisons. The last review of the text, outside of the publishing house, was made by Hugh Iltis, professor emeritus at the University of Wisconsin, who offered abundant corrections and changes. Yes, I heeded them, the most memorable being rewriting the title of a chapter from merely "Speaking in Academia" to "Speaking, Teaching, and Raising a Little Hell in Academia."

The idea of including photographs was suggested by Fjaere Nilssen, my wife's daughter and a singer-songwriter in Los Angeles. Mack Prichard, my buddy on (and off) the trail, provided many of the photographs and helped in selecting others. I received utterly essential computer aid from Richard Navas, Steve Cheek, and Sonya Hess.

I recall that I was already working on *Greenspeak* when I was privileged to write the foreword for a new edition of *Out Under the Sky of the Great Smokies* by my late mentor and friend, Harvey Broome. That went very well, leading Scot Danforth of the University of Tennessee Press to express the hope that we could do something together again. So here now we have done it, with Scot helpful, cooperative, and supportive all the way. And so too Karin Kaufman, the considerate and creative copyeditor, who rescued me from a few boo-boos, eliminated a few repetitions, and tolerated leaving two or three others in place.

Finally, I acknowledge the essential contribution of my sweetheart and wife, June Eastvold. We pursue parallel interests, believe in each other, and lead each other's cheering section.

Sherman Adams

At the annual dinner of the Society for the
Protection of New Hampshire Forests,
Shelburne, New Hampshire,
September 24, 1971.

Mike Frome gets no biblical references from me; he is too well qualified to quote his own. Though the name somewhat fits him, I shall not call him "the Ralph Nader of resource conservation," for he knows a lot more about his subject than the nemesis of General Motors.

Frome is an editorial writer, author and commentator, and professor emeritus of introspection. Sometimes his work shows the cutting edge of the educated critic, and often the constructive offensive of the crusader who searches for better answers to the myriad problems confronting the conservationist who does battle with the overuse, misuse, and abuse of forest resources.

An air transport navigator in World War II, decorated for his many overseas missions, he has much later navigated his daughter into Williams College, a former male sanctuary, a maneuver, incidentally, which deserves no decoration. *O tempora, O mores,* what bastions are left to the infiltrated male, laid siege in the guise of women's lib!

All of which is unrelated to the fact that Mike Frome is a well-founded newspaperman who is about to speak his mind as he sees fit, and you foresters better repair to your redoubts, or throw up a few breastworks, for here comes Mike Frome.

Sherman Adams, trained as a professional forester, became a congressman, governor of New Hampshire, and chief of staff for President Dwight D. Eisenhower.

Jim Robey

At the annual meeting of Little Miami, Inc.,
Yellow Springs, Ohio,
March 19, 1973.

A week ago I had occasion to refer to the *Audubon Encyclopedia,* specifically to the section on wildflowers. It begins with a quote from John Burroughs: "If we think birds, we shall see birds wherever we go; if we think arrowheads, as Thoreau did, we shall pick up the wild flowers in every field."

Were Burroughs alive today, he might have added: "If we think environment, as Mike Frome does, we shall have a better environment."

I know of no writer who thinks more about the environment and writes so diligently about it as Mike Frome.

Through the eight books he has written, scores of magazine articles, talks to groups such as ours, and his monthly conservation essay in *Field & Stream,* Mike has alerted millions to environmental issues.

In World War II he was an Air Force officer, but I would have guessed he was a Marine. In the conservation battles, he has been the first on the beach.

Before it was a fashionable topic, Mike was warning us about how our national and state parks are being overrun, about the congested camping and the desecration from recreation developments.

He has always been a thorn in the side of the timber industry for abusive overcutting and underplanting. He has kept a stern eye on the U.S. Department of Agriculture for its cutting practices in the national forests. In fact, Mike has done such a good job he was fired as a writer for the American Forestry Association magazine, which only endeared him more with those who really do want a better environment.

The first time I heard the words "Alaska Pipeline" was nine years ago. It was Mike Frome who sounded the warning at the North American Wildlife and Natural Resources Conference.

I met him six years ago on another river—the Tennessee. With the welcome he was given in Chattanooga you would have thought that— single-handedly—he saved the Little Tennessee from the Tennessee Valley Authority. He and Mrs. Frome were presented with a beautiful

---

Jim Robey was outdoors editor of the *Dayton (Ohio) Daily News.*

painting, and they were given a cruise down the Big Tennessee on one of the largest boats ever to cruise on the river. The rest of us writers had to get down anyway we could—even if it meant rowing.

For all that Mike Frome has done for the environment, for his desire to help us save our river, give him a welcome he'll never forget!

Sam Ham

At the annual Northwest Regional Conference,
National Association for Interpretation, Moscow,
Idaho, October 20, 1984.

In conservation history, we celebrate the impacts that visionaries such as George Perkins Marsh, Henry David Thoreau, John Burroughs, Frederick Law Olmsted, John Muir, and others have had on American thinking about nature, conservation, and the environment. These are what we might call the "heavyweights" of conservation writing. And it is my firm belief that our speaker tonight falls into that category with them.

When I asked how he would like to be introduced, all he said was, "Be sure to tell them I'm your friend!" I'm sure I'm honored, but Michael is a friend to everyone. More than that, Michael Frome is a friend of wild nature, and his writings and public addresses reflect more of a love affair than a casual friendship. He has been and continues to be one of the clearest and most heeded voices for wilderness in the world—and we are indeed fortunate to have him as our speaker tonight.

About a year and a half ago, Michael joined the Wildland Recreation Management faculty at the University of Idaho as a visiting associate professor. The night he arrived in Moscow, I had him to my house for a welcoming dinner. And although I had read many of his books and followed his career, I learned things about him that night far more personal and much more indicative of the type of human being he is.

I learned that Michael loves children and he is intent on leaving a wonderful planet for them and their children. I learned that Michael is humble despite the fact that he has been the personal acquaintance of every natural resource bureau chief since Newton B. Drury's time at the National Park Service, despite his fame as an author and columnist, and

---

Sam Houston Ham is a professor at the College of Natural Resources, University of Idaho.

despite the fact that he is known to eat lunch at the Cosmos Club during his frequent stays in Washington, D.C.

I also learned that Michael is honest. In fact, he's as frank and straightforward as they come. Impressed with his sometimes piercing journalistic candor, Interior Secretary Walter Hickel remarked, "I consider Michael Frome one of he finest environmental writers in the nation—courageous, accurate, and widely respected for his integrity. Mike tells it like it is, not necessarily like we'd like to think it is."

And has he ever told us. It seems every time Michael puts pen to paper—or more frequently, fingers to keyboard—something poignant and significant results. And the world looks forward to more Michael Frome books currently in preparation. Michael's topic tonight is censorship. Titled "To Sin by Silence: Censorship and Interpretation," the focus of his presentation seems not only important, but timely in this age when our audiences not only have a right but a desire to know about the fate of their natural heritage. This will be another case where Michael "tells it as he sees it."

Nathaniel P. Reed

At the Vail Symposium, commemorating the seventy-
fifth anniversary of the National Park Service,
Vail, Colorado,
October 10, 1991.

I have known Michael Frome for many years. More important, I have read Michael Frome's hard-hitting analysis of the growing problems in the federal land-management agencies. Michael is tough, uncompromising. He is angry—angry at stupidity, angry at complacency, angry about malfeasance, misfeasance, and, worse, nonfeasance. To have the authority to act and sit on one's hands is just as much a crime as larceny.

Michael likes to stir things up. His basic credo is to ask "Why?" Many "whys" are painful. If a program has been run on an even course for many years, why should any manager look forward to being asked "Why?"

Nathaniel P. Reed of Florida served as assistant secretary of the interior under Presidents Richard M. Nixon and Gerald R. Ford, and at various times on boards of directors of the National Audubon Society, National Geographic Society, and National Parks and Conservation Association.

Michael is restless with the status quo. He is the champion of the energetic, the inquiring, the bold, the innovator, and the questioner.

Not since the days of the robber barons of the nineteenth century have our nation's natural resources been so badly managed, or morale of our land managers been at such low ebb. This is the timely moment when Mike Frome's clarion call is being heard, urging the young to be bold and the old to stand their ground and preserve their honor.

This is a role that makes few friends in high places, but is a role that is desperately needed and must be heeded.

Take the Bull by the Tail
and Get a Toehold

I never considered myself a spellbinder. I rarely brought slides or over-heads. I did not want to become a professional speaker, believing that successful speakers become actors, rehearsing, memorizing, responding on cue, and doing it for money. I tried not to give the same speech twice, and whenever I did found it boring and really shortchanging to whomever came to hear me.

The invitations to speak to the best of my reckoning came in particular periods: following publication of a new book, while I was conservation editor of *Field & Stream,* when I was fired by *Field & Stream,* and when I was working as a university professor. Celebrity, notoriety, and authority, in whatever dimensions, seem to be factors in making a speaker attractive.

In 1975 I went south to speak at the dedication of the Sipsey Wilderness, a chain of deep gorges threaded with streams and waterfalls which the Alabama Conservancy had labored diligently to protect. Before the ceremony and speeches, I was signing copies of *Battle for the Wilderness,* then lately published. A young woman who worked for the Alabama Conservancy and thought I was some kind of celebrity hung at my elbow. Finally I told her I needed time alone to work on notes for my speech. "No you don't," she said. "Get up on the platform and speak from the heart."

I've thought of her words many times. Yes, a speaker ought to know his subject well enough and be composed enough to speak from notes stored in the head and heart. But it isn't that simple to be fully effective. My wife, June Eastvold, graduated from the University of California, at Berkeley, with a degree in rhetoric, which helped prepare her for her career in ministry, including sermonizing every Sunday over a period of years. I've watched and listened. She works hard, researching and writing the complete sermon in advance, then delivering it on Sunday as though every word is fresh and new. The congregation always stays awake and responds positively to her message and delivery, and to her warmth.

In contrast, I've heard environmental speakers who sounded didactic, like commissars. They dwelled on how bad things were and dressed down instead of up as if to prove it. I remember a particular evening program in Seattle when the Northwest Ecosystem Alliance introduced a new book with commentaries by several of its contributing writers, all downbeat, gloomy, unsmiling, using big, technical words in every sentence.

I never regarded myself as a great speaker, and I'm sure there's plenty of room for criticism of my delivery and message, but I tried to make it with enthusiasm and hope. For better or worse, I injected jokes and stories, which I collected here and there. Like the congressman from California who said to the congressman from Colorado, "What we have to do is take the bull by the tail and look the situation squarely in the face." To which his friend replied, "You're absolutely right. We must get a toehold in the public eye!"

I particularly favored fables about writers—Samuel Johnson, for instance, who was badgered by an ambitious young fellow asking the great one to critique his draft essay. Finally Johnson agreed; he reviewed and returned the opus. The writer eagerly opened the envelope. "I have read your manuscript. It is both good and original," Johnson's critique began. Oh, wow! But then it continued, "The trouble is the part that's good is not original, and the part that's original is not good."

Or try this one: I was fishing in the Smokies when I met a fellow along the creek. He was fishing too, and we got to talk. After a time he referred to the hardware store he and his brother owned in town. Then he asked, "What do you do, friend?" When I said I was a writer, he snapped back, "Yes, but what do you do for a living?"

Or try this: A newspaper reporter lost his job and tried to make it freelancing. He sold a few pieces, made a little money and then a little more. One evening he took his wife to dinner and announced they now could afford a house cleaner twice a month. So Alice was hired. She came and for a time all went well. However, in due course she approached the lady of the house.

"Ma'am I've been coming here to do your work for three months," ventured Alice.

"Yes, Alice, and you've been a big help. What is it?"

"Ma'am, you've been paying me twenty-five dollars . . ."

"Oh, I see. But I can't afford more just now."

"Oh, it isn't that." And then Alice paused, looking over her shoulder to the study where the unshaven man in sweatshirt and sneakers had his feet on the desk and head buried in a book. "It isn't that at all. I just want you to know I'm willing to take less until your husband can get a job."

Probably the best stories, funny or otherwise, related to the gospel of citizen empowerment, the belief that laws and the democratic process were meant to work in the public behalf. Occasionally I would mention that I had been in Iceland in 1972 at the very time Bobby Fischer, the irreverent American chess player, met and defeated Boris Spassky, the Russian master who was then world champion. Fischer had a saying: "All that matters on the chessboard is good moves." He evidently snubbed the upper echelons of Iceland society but made friends with everyday folks. On the last day of the chess match with Spassky the fishing war with England began. The British continued netting cod inside the fifty-mile limit Iceland protected for its own fishermen. "Now the next move," Fischer counseled his friends, "is to push the limit to two hundred miles."

To make the point in another way: The grass-roots citizens group tried everything in its campaign in the home community. Finally they decided to go lobby in Washington. Just when they had an appointment with their silver-haired senator, he was called to the floor. He prided himself on his ability as an orator and invited the delegation to hear him from the gallery. After delivering what he considered a particularly fine speech, he met with the delegation. A little lady in tennis shoes exclaimed breathlessly, "Senator, that speech was absolutely—yes, it was absolutely, absolutely superfluous!" The senator was taken aback but recovered quickly. "Thanks very much. I am glad you enjoyed it. As a matter of fact, I am planning to have it published posthumously." Whereupon the little old lady said, "Wonderful, wonderful—the sooner the better!"

Speaking and writing differ considerably, though both are media of communication based on words. The speaker is able to establish eye contact with a live audience, to emphasize points with pauses, gestures, body language, whispers, and shouts. A pastor composes sermons to be spoken rather than to be read. On the other hand, many writers feel a lot freer, and maybe safer, when creating images on paper without personal contact or the need to perform. For better or worse, through my career I plunged onward, making the best of whatever opportunities came along. By going out to speak I thought I could test ideas and listen to people and learn from their ideas.

But speaking in the basic sense is like writing. It pays to focus on a particular point or theme, to develop it fully with forethought and preparation. Whenever I did my homework I knew I could relax and avoid shuffling papers at the podium. I learned to arrive early enough to get acquainted with people and the room and to test the audio system.

A good speaker ought also to be a good listener. I've been deeply touched and influenced by certain speeches I've heard and will cite three of them here.

One was delivered by Sigurd Olson when he was awarded the John Muir Medal at the 1967 Sierra Club Wilderness Conference in San Francisco. Olson was prominent as an author and activist working in his home country, the Boundary Waters Wilderness of northern Minnesota. He was afflicted with Parkinson's disease, or palsy, and usually shook badly, but when he spoke he became firm and steady so the audience would never know. At the San Francisco conference he delivered a strong, eloquent message, with presence and poise, challenging believers in wilderness, the already converted:

> The stakes are so high, the threat so desperate, we can no longer think of wilderness as being a minority need, a need of 2 percent of the population. I feel that the wilderness is the concern of all Americans and all humanity, that if we do not save some wilderness, mankind and his spirit will suffer, and life will not be so happy for future generations.
>
> My only suggestion to this conference is to consider, as wilderness battlers, ways and means not for reaching each other—we are converted—but for reaching the other 98 percent of the people. Make the wilderness so important, so understandable, so clearly seen as vital to human happiness that it cannot be relegated to an insubstantial minority. If it affects everyone—and I believe it does—then we must find out how to tell the world why it affects everyone. Only when we put wilderness on that broad base will we have a good chance of saving it.

I felt that he was speaking directly to me, in an uplifting manner that I could not avoid, and have tried to respond to his call ever since.

Another memorable speech in my life was delivered by Marjory Stoneman Douglas in 1985. She had come to Washington, D.C., for a ceremony conducted by the National Parks and Conservation Association to present the Marjory Stoneman Douglas Award to the "national parks citizen of the year." That award was named for her in recognition of her writing and activism, which reshaped the public image of the Florida Everglades. With poetry in her prose and spiritual power, she showed that the Everglades is not a swamp to be dredged, filled, and converted into real estate but a rare ecological wonderland. Since its publication in 1947, her book, *The Everglades: River of Grass,* has projected an image of hope and heart, for the Everglades and all endangered fragments of God's earth.

At the awards ceremony in Washington, I was the recipient but she was the star of the show. She was ninety-six years old, nearly blind and mostly deaf, yet vigorous and living in the future rather than the past, with a message of determined optimism, sternly warning politicians and public officials that if they didn't act forthwith to save the Florida panther she would get school children all over America to write letters. In 1991, she was still going strong, celebrating her hundredth birthday at a Miami nature center named for her with a pledge to keep fighting to save her beloved Florida. And so she did until she died in 1998 at age 108.

Then there was William O. Douglas on that day in May 1977 when the Chesapeake and Ohio National Historical Park was dedicated to him "in recognition of his long outstanding service as an American conservationist and for his efforts to preserve and protect the canal and towpath." He had suffered a stroke and retired from the Supreme Court. Now it was time for the ceremony beside the canal in Georgetown, the oldest part of Washington. The chief justice and almost all associate justices of the Supreme Court were there.

The crowd stood and cheered as Douglas was wheeled in. I felt it was a deserved salute to one of the truly great Americans of my lifetime. Conservation was only one of his concerns. As a strong-willed civil libertarian, Douglas bucked the political or public tide when he felt need to. "We must have freedom of speech for all," he insisted, "or we will in the long run have it for none but the cringing and craven." Richard Nixon and his vice president, Spiro Agnew, and the Republican leader in the House of Representatives, Gerald Ford, tried to oust him from the Supreme Court, but he brushed them off. It grieves me that in Washington state, which Douglas called home and where I live, I can rarely find a college student who knows any more about him than his name, maybe not even that.

At the ceremony in Georgetown, I was shocked (as I'm sure others were too) to see Douglas frail, ashen, his face contorted with partial paralysis. His lovely wife Cathy prepared to read his remarks, but he insisted on speaking for himself. He paused between sentences in an unnatural cadence, but the clarity of his mind was unquestionable. He reminisced about Justice Brandeis, who used to canoe on the canal, and then thanked everyone for coming: "I thank all those who have no portfolio but who have two strong legs and like to hike, and to listen to the pileated woodpecker. I promise to get well, and we'll be able to walk in again."

He never made it again, but William O. Douglas personified courage and independence. From the mid-fifties until the seventies he responded to calls from people in Kentucky, Tennessee, Arkansas, Maine, Wyoming, and Texas. He visited particular areas they were trying to save, spoke to their meetings, and gave encouragement around the campfires.

Sometimes I went where he had already been, picking up his gospel of citizen empowerment, a belief that laws and the democratic process were meant to work on behalf of the public. Sigurd Olson and Marjory Stoneman Douglas, those other heroes of mine, were much the same, apostles of greenspeak.

Where I Came In

In the late 1940s I drifted from newspaper work into public relations for the American Automobile Association. Many times in retrospect I have considered that move the worst career mistake in my life, although ultimately it led me to where I am, with fulfillment in an altogether different writing and activist career. Besides, at the AAA I met and married my first wife. We had two children, our progeny, and stayed together thirty-two years, until my belated midlife need to change led elsewhere.

The national headquarters of the American Automobile Association were located in Washington, at Seventeenth Street and Pennsylvania Avenue, one block from the White House, but the AAA was really a federation of affiliates located all around the country. They were called "clubs," dating from the days early in this century when motoring was a sport and adventure and when civic groups campaigned for better roads and for acceptance of the automobile as successor to horse and buggy. Some clubs in my time were still civic bodies, motivated by public service as they saw it. Others, however, were privately controlled and corporate driven.

All of them taken together constituted big business, selling insurance, road service, tours, and cruises, and constitute a much bigger business today. Attempts have been made now and then to expose the American Automobile Association as crass and commercial, but I don't think it's worth the effort. In my time membership reached four million; in the year 2000 it exceeded forty million. "Subscriber" might be a better term than "member," but I'm one myself.

When I went to work there, in post–World War II, everything in America was in accelerated change. The war boom continued so that people had jobs, earned money, bought cars, went places on paid vacations over new highways, stayed in new motels (complete with toilet in every room) that replaced the old jerrybuilt roadside cabins, and discovered Yellowstone, Yosemite, the Grand Canyon, and other national parks.

As part of my work, I became a bit of an expert about all of this. I was called the "travel editor." I tried to keep abreast and wrote articles and made speeches for the AAA. Here, for example, is the report of one as it appeared in the *Miami Daily News* on April 9, 1952:

Editor Expects Record Florida Summer Season

Tourist travel this spring and summer will establish an all-time record, Michael Frome, travel editor of the American Automobile Association, predicted today in Miami Beach.

Frome spoke before about 150 Greater Miami businessmen at the first annual Spring–Summer travel seminar at the Sans Souci Hotel.

Information compiled by AAA offices throughout the country indicates that about 15 per cent more motorists intend to visit Florida this year than did in last year's record-breaking season, Frome said.

The following month I spoke in Asheville, North Carolina, predicting a 20 percent increase over last year's all-time record, which tourist promoters there loved to hear. My files remind me that I spoke in Mississippi, the Ozarks, and Reno, Nevada. I traveled to a lot of places and wrote articles about them. In due course I became involved in civic programs that extended my horizons and introduced me to new people. For example, the AAA was part of a national anti-billboard campaign and of efforts to protect roadsides from commercial "ribbon development." Those were good causes that I wish had proven more successful. I went to national parks and got to know officials of the National Park Service, at the agency's headquarters and in the field.

Thus it happened that in 1952 I accompanied the president of the AAA, Ralph Thomas, a Michigan businessman, when he went to address the Virginia Travel Council in Roanoke, his hometown. While there we were taken on a tour by Sam P. Weems, superintendent of the Blue Ridge Parkway, whom I already knew as a friend. In my mind the Blue Ridge Parkway, a part of the National Park System, was a model for recreational motoring, extending 469 miles along the mountain spine of Virginia and North Carolina free of truck traffic and billboards, with access to parks, forests, and mountain vistas along the way. However, many miles of the parkway were still unbuilt, and Weems endeavored to enlist support from Thomas, and through him the AAA, for increased federal funding for accelerated construction. Once back in Washington I thought this little episode would be forgotten, but a week later the executive vice president of the AAA, Russell Singer, called me in and directed me to prepare a statement of support as Thomas had wished. At the time I had no way of knowing what that would involve and where it would lead.

I thought it would be easy. I would telephone Herbert Evison, the chief of information of the Park Service, for data on the Blue Ridge Parkway, which I would edit or rewrite into a letter for Thomas. But it didn't work that way. "It isn't only the Blue Ridge Parkway that needs help, it's the whole park system," Evison said. He asked me to come by, which amounted to a short three-block walk from the AAA to Park Service offices in the Interior Department. I felt confidence in Evison, who came across as a principled parks person rather than a public relations man. He explained that national parks had been virtually closed during the war but now were being deluged by millions of visitors, even though facilities and funding were at inadequate prewar levels. Although I was unaware of it at the time, Bernard DeVoto, the historian and essayist, was working on what proved a classic piece on this subject titled "Let's Close the National Parks" in *Harper's* October 1953 issue.

On February 3, 1953, Ralph Thomas sent a letter (that I had drafted) to appropriations committees of Congress. He reviewed the interest of AAA clubs and their members since 1916, when the National Park Service was established and automobiles were first allowed in the parks. He wrote, "These areas are approaching a time of crisis; unless action is taken soon to provide adequate protection and new facilities, there is danger of lasting and irreparable damage. . . . We are mindful of the fiscal responsibilities of the Congress. But we do urge that Congress recognize the basic needs of the National Park Service to maintain its equipment and property and to prevent further deterioration, and to provide adequate facilities for protection and reasonable accommodations for the growing army of visitors and to meet necessary sanitary conditions."

This proved to be only the first step. I had more discussions with park people and internally at the AAA. I carpooled for a time (from my home in suburban Virginia) with Ronald Lee, an associate director of the

National Park Service, and focused at length about what more the AAA could do for the parks. This led to a dinner and dialogue hosted by the AAA on December 10, 1953, at the Metropolitan Club, an elite men's club in Washington attended by the following: Douglas McKay, secretary of the interior; Melville Bell Grosvenor, editor of *National Geographic Magazine;* Laurance Rockefeller, the philanthropist in conservation; two former senators, Gerald P. Nye of South Dakota and Burton K. Wheeler of Montana; Conrad Wirth, director, and Ronald Lee, associate director, of the National Park Service; and for the AAA, Andrew J. Sordoni of Pennsylvania, president (having succeeded Ralph Thomas), Russell Singer, executive vice president, and myself.

Wirth at the dinner distributed a brochure, *The National Park System—Present and Future,* prepared for the occasion. He used it as the basis of discussion on how to win friends and influence public policy on behalf of the national parks. It was one of the early building blocks that led to the ten-year national park development program to be called Mission 66.

Subsequently, the AAA and secretary of the interior cosponsored in 1954 and 1956 what we called American Pioneer Dinners. The first was held at a downtown hotel. The second, at the Interior Department, was attended by sixty members of the Senate and House and two Supreme Court justices. The state of South Dakota sent buffalo and elk meat for the main course. Ansel Adams, the celebrated photographer, sent prints of his work in Yosemite for everyone present. After dinner, guests went upstairs to the department auditorium to see a film, *Adventures in the National Parks,* prepared especially for the occasion by Walt Disney. And the National Park Service distributed the booklet *Our Heritage,* the first official presentation of Conrad Wirth's brainchild, Mission 66.

I was in the middle of it all. On February 23, 1954, Wirth wrote me:

> You, Ronnie [Lee] and John Doerr [another Park Service official] and the rest made this dinner possible. To you should go the greater part of the credit for making it the success that it was. It was a tremendous chore, and the constant changing of dates and whatnot I know was a tremendous strain on you. However, I cannot think of anything that has been better received, or more successfully staged, in my twenty-six years in the Government, than the Pioneer Dinner.

The reaction on the Hill and among our friends of
the press will be everlasting; they will be talking about it
for years. I am sure that it will have the desired results
of bringing to the attention of the legislative and administrative branches of the Government, and the public,
the need for close working relationship and understanding of the problems of the National Park Service.

Mike, I thank you for everything you have done
from the bottom of my heart, not only for myself but
on behalf of the entire Service.

He wrote me a similar letter February 16, 1956, after the second dinner: "Without your fund of ideas plus your tireless efforts in planning
and executing the many details involved, the affair could not have been
the great success that it was."

Maybe so, but now I look back in retrospect at what I've learned and
how I've changed. In his memoir, *Parks, Politics, and the People,* Wirth
wrote that officials of conservation groups and others influential in the
conservation field were invited and welcomed. Devereux Butcher, however, was not among those present, nor was he on the invitation list. I
remember that he called four or five times, explaining that he was with
the National Parks Association (later, the National Parks and Conservation Association) and that he was very interested and wanted to come.
Since I didn't know him and knew little about his organization, I consulted Ronald Lee, my collaborator at the National Park Service. He
urged me *not* to invite Butcher because to the Park Service he was a
problem and a bit of a pest. I dodged Butcher's phone calls and ducked
when I had to speak to him. In due course I recognized him as a true
believer in national parks and was sorry.

Devereux Butcher worked for the National Parks Association from
1941 to 1959, variously as executive secretary and editor of its journal.
He fought to protect the parks as the sanctuaries they were meant to be.
Years later we corresponded and he provided valuable data. He wrote
me May 10, 1987, a year before he died, deploring the intrusion of cruise
ships in Glacier Bay National Park, elk shooting in Grand Teton National
Park, hang-gliding, downhill ski resorts, and snowmobiles in other parks.
He sent me a copy of the eighth edition of his book *Exploring Our
National Parks and Monuments.* In the preface he wrote:

Since the parks are nature sanctuaries to be held intact for all time, there must be no activity within them such as mining, logging, dam building, airport construction, and grazing; nor conforming, crowd-attracting facilities such as golf courses, swimming pools, tramways, ski lifts, tennis courts, dance halls, nor pastimes such as hang-gliding and snowmobiling. . . . Like literature, music and art in their highest forms, they contribute to our spiritual well being, and they require unending vigilance to preserve them for that purpose.

I could not agree more and wish that others, in policy positions, would too.

As for Mission 66, Conrad Wirth's legacy, Congress did indeed boost appropriations as he had hoped. Considerable funding went to purchase private holdings within park boundaries and returning them to nature. That was good. Important new park units were established, including Cape Cod in Massachusetts, Padre Island in Texas, and Point Reyes in California. That was good too. Mission 66 brought a lot of building and construction of roads, parking areas, utility systems, marinas, visitor centers, and training centers. Some parts of it were good and other parts questionable, but in my opinion, higher priority was placed on visitor comforts, facilities, and enjoyment than on protection of natural systems.

At the outset of Mission 66 Wirth said he hoped to stay as director until its completion in 1966. Yet he turned in his letter of resignation to Secretary of the Interior Stewart L. Udall much earlier, on October 18, 1963. Wirth wrote Udall that the decision to leave was all his idea, that the secretary did not want him to retire but "you indicated that you would respect my wishes." A United Press International dispatch dated October 29, 1963, reported, "Wirth's decision to retire was his own and was reached long before the announcement last week, Udall said in an interview. The Secretary said he was 'appalled' at reports the resignation was forced. . . . 'I'm saying flatly there was no lack of confidence [in Wirth] at any time.'" Maybe so, but when I interviewed Udall in 1985 he told me that he had long wanted to get rid of Wirth, whom he had considered too old and out of step, and that when, in the course of an argument, Wirth had threatened to quit, Udall exclaimed forthwith, "I accept your resignation!"

I observed a lot of those games of fictions and fables. I like to believe that I never played them, although I certainly made mistakes and said things that I regret. For example, in 1957, one year before leaving the AAA, I spoke at the annual meeting of the National Trust for Historic Preservation at Swampscott, Massachusetts. John Fenton of the *New York Times* reported that the meeting focused on the impact of the federal highway program on historic structures. The concern was over how much to save and how to save it:

> Throughout the three-day session, speakers continually stressed the premise that mere antiquity was not sufficient basis for selection of a project for permanent preservation. Antiquity could be a factor, it was pointed out, if other more significant examples had disappeared or if a particular building was especially characteristic of a given community.
>
> The underlying theme of the conference was clearly delineated by Michael Frome, travel editor of the American Automobile Association, in a discussion of the likely impact of the Federal highway bill of 1956.
>
> Mr. Frome noted that as Americans had taken increasingly to wheels since World War II, there had been a corresponding rise in appreciation of natural scenery for its own sake.
>
> The automobile association editor, who travels thousands of miles each year, asserted that by bringing hitherto inaccessible historic sites within reach of millions of persons, roads should be regarded as an accessory rather than a deterrent to historic preservation.
>
> Mr. Frome said that roads, as a matter of history, had helped create the American heritage. He cited the Boston Post Road over which the Colonial mails were carried between New England and New York. Others included Braddock's Road, the Wilderness Trail, the Natchez Trace, the Chisholm Trail and Santa Fe Trail.
>
> It was the policy of the United States Bureau of Public Roads, said Mr. Frome, to weigh carefully scenic and historic values that might be placed in jeopardy. Thus, since this policy included the holding of public hearings

on proposed routes, historical societies and preservation groups had a responsibility to point out where dangers were involved and to provide "mature counsel" to
ensure the saving of sites because they were worth saving "and not simply because they are old."

I was wrong, to the nth degree. The old roads may have played their
part in history, but the modern road builders have done incredible damage to historic and natural values. I wish that I had sounded the alarm to
those people engaged in saving the heritage.

At least I was trying. Friends who were travel editors increasingly
asked me to contribute articles about parks and preservation. My early
articles were published in the *Christian Science Monitor* and *New York
Herald Tribune* and, later, after I became a freelance writer, in the *New
York Times* and *Chicago Tribune* until ultimately the *Los Angeles Times*
gave me a column of my own, "Environmental Trails," for six years.
Maybe I was becoming too much of a personality for the American Automobile Association, but I was dismissed unceremoniously in 1958, the
only explanation being that it was time for a change. Though it hurt at
the moment, it proved a blessing. It was time to move on, to clear my
head and raise my sights.

Soon after leaving the AAA I read an article in the *New York Times*
travel section of February 22, 1959, titled "Invasion of Baja California." It
was not written by a travel writer but by Joseph Wood Krutch, a well-
known literary figure of that time. I clipped the article and still have it in
my files. The article was datelined LaPaz (capital of the state of Baja California Sur) and began, "The long, ruggedly beautiful peninsula called
Baja (or Lower) California came early into Western history but has stayed
pretty persistently out of it ever since. Now, after more than four and a
quarter centuries of stubborn resistance to everything called progress, it
has begun to undergo what might be cautiously called the beginnings of
a tourist boomlet."

Krutch described the "mixed blessings" to "one of the most nearly
untouched areas of scenic beauty still left on the American continent."
He concluded with a scene in the tropical village of San Bartolo, midway
between LaPaz and the cape (at Cabo San Lucas):

> There is one store and an old stone chapel.
> Surely—although I did not happen to see him—there
> is also a guitarist. But as I stood by the spring watching

the women carrying water, they did not appear to feel put upon. As for the vaquero loitering there on his horse, a good deal more could be said. Held erect in the saddle by the last vestige of Spanish pride, he was certainly not longing for the tourist trade. His son may very well get it whether he wants it or not—but it is not certain he will be better off.

I was impressed by Krutch's theme and style in this article, and by his life. In 1952 he had quit New York for Arizona and there became a worthy chronicler of the desert and desert life. "By contact with the living nature we are reminded of the mysterious, nonmechanical aspects of the living organism," he wrote. "By such contact we begin to get, even in contemplating nature's lowest forms, a sense of the mystery, the independence, the unpredictableness of the living as opposed to the mechanical." He was a conservationist in his own way and undoubtedly influenced me to be one in my own way too.

Recreation—A Substitute for Conservation?

At the Fifth American Forest Congress,
Washington, D.C.,
October 29, 1963.

In 1962 my book *Whose Woods These Are: The Story of the National Forests* was well received and well reviewed and I was invited to speak here and there, principally in forestry circles. In that period I was close to the American Forestry Association, a respected old-line organization, and later for six years wrote a column of opinion in its monthly journal, *American Forests*. The AFA was headquartered in an attractive old building facing a park near the White House, with a cross-section of a redwood log out front that enhanced its image in Washington.

The AFA called itself the country's oldest conservation organization with some rationale. It was established in 1875 under the auspices of the American Association for the Advancement of Science at a time when large areas of forest were being dominated by corporate control and devastated by clearcutting and fires. The first American Forest Congress, conducted under its auspices in 1882, led to a pioneering study by the federal government of the country's forestry needs. The second congress, in 1905, was addressed by President Theodore Roosevelt and led to establishment of the United States Forest Service as the agency designated to administer the forest reserves, henceforth to be called national forests.

The association was at the conservative end of the conservation movement. It considered itself all-embracing, though most of its directors came from the timber industry, Forest Service, and forestry schools, with recognition also for private woodlot owners. These interests were well represented at the fifth American Forest Congress, as were the

American Mining Congress, United States Chamber of Commerce, and the chemicals industry. I spoke on the panel on recreation, along with the director of the Bureau of Outdoor Recreation (an agency of the Interior Department established in the Kennedy administration and eliminated in the Reagan administration), the executive director of the American Forest Products Industries, and the executive director of the Nature Conservancy, so there was a breadth of viewpoints.

In later years the association declined in influence and membership. It sold its building, moved to a less prominent location, and changed its name to American Forests.

Conrad L. Wirth had lately resigned as director of the National Park Service under circumstances that indicated political intrusion into a professional agency. The conference organizers asked me to discuss this point. In general, I was learning but still had a lot to learn.

■ ■ ■ ■

Ten years ago this December [i.e., 1953] I attended the Mid-Century Conference on Resources for the Future, held here in Washington, and participated in a group considering rural land uses. I recall that recreation was one of the lesser items on the agenda. For every other point on the program, one person had been designated as the discussion leader. But there was none for recreation, and so I offered to speak first, urging more funds for the national parks, greater recognition of the recreational aspects of the national forests, and more emphasis by state and local governments on the development of recreation and vacation areas to keep pace with the needs of the American people.

Conservationists have traveled a long road in the past ten years. Recreation, the forgotten tail ten years ago, has moved virtually to the head of any discussion on land use. But I like to think that I have traveled a long road, too, and have learned that merely by meeting recreational needs alone this country cannot fulfill its obligations toward its natural resources.

It seems to me that in many ways conservation is fast becoming the tail of the recreation giant. Chair lifts, resorts, roads, campgrounds, and reservoirs are increasingly bestowed upon us. But I do believe that these times—when the precious land is vanishing, and with it the birds and animals, the clear streams and wild places, which are part of the national soul—demand that all of us, who constitute the great outpouring of recreational use, must endeavor to understand the meaning of conservation in its broadest sense.

We should not anticipate or demand, or tolerate, the transformation of all our open space into one vast playground for motorcars, motorscooters, and motorboats. We should not run roughshod over other uses, whether logging, mining, grazing, watershed or wilderness, but try to find the means whereby all of these are planned for as complementary units of a harmonious whole. Nor should this be the responsibility of only political leaders or land managers. The fishermen, canoers, campers, boaters, bird watchers, skiers, hikers, and hunters must do more to demonstrate responsibility to the land than to assert our inalienable rights to use it. We must participate, in the same sense that we participate in a parent-teacher association, in the affairs of a hometown, in the management of a business, for the management of the land is everybody's business. The tragedy is that we know more about outer space than we do about our own resources, that we are much better informed on the situation in Vietnam than about timber supplies and the water crisis, which are basic to our survival. We simply do not know, we are in the dark on conservation, we are permitting ourselves to be deluded into accepting recreation, complete with blatant billboards, resort living, and mechanized convenience, as a substitute for conservation.

This is why I am here before you, who are concerned with the course of conservation, to emphasize that your efforts depend not on the mass of recreational facilities you provide, of whatever nature, but on the extent to which the users are enlightened and understand their fundamental stake.

For this outlook on life, I am indebted to the association sponsoring this congress. A few years ago, while trail riding in the high Rockies as one of a group conducted by the American Forestry Association, I felt overcome by the marvels of the raw American earth that surrounded me. The fact was that I had had nothing to do with placing or arranging them as I had found them. But in my hands, and in the hands of others like me, lay their fate and future. One might say that in the instant of truth, out where the sky and mountains meet, this recreationist experienced the meaning of the phrase "Think not of what your country can do for you, but of what you can do for your country." On that day of reckoning, I decided it was my obligation to find out how these lands are managed, and by whom, and why, and in what direction for the future.

The core of the answers to these questions is not easy to come by. We look to the president [John F. Kennedy], who recently embarked on an expedition to bring the story of conservation before the people. It began on a marvelously high plane at the old Gifford Pinchot estate in Pennsylvania, where his words stirred the conscience of the country to appreciate the deep meaning of our natural resources and the need for their protection. But as the president made his way westward it became plain that he was not speaking of conservation in precisely this sense. It was recreation that he meant, the kind of play and pleasure enjoyed by large numbers of people, and the expenditure of large amounts of federal funds for dams, reservoirs, and big power projects. By the time he was through, it was universally evident that he had missed a magnificent opportunity to present conservation as a philosophy, a belief, a way of life that stands above politics, deserving the support of all.

We look to our lawmakers in Congress, many of whom have awakened to the recreational surge, not necessarily to spur appreciation of the force of conservation, but to satiate the appetite for outdoor playgrounds. I read a speech recently by the chairman of the Interior Committee of the House of Representatives, Congressman Wayne Aspinall, on the "Recreational Aspects of Conservation." Said he, "Congress is and should be a stabilizing influence in developing a conservation policy." Though I hope this shall be the case, sometimes I wonder, while one member after another campaigns to establish a national park area in his district, though such are often clearly below the standards of the "jewels of America," or one member after another intercedes for chair lifts, resorts, and other intrusions advocated by the local chamber of commerce. Sometime I would like to hear Congressman Aspinall, or some other prominent figure in government, speak on "Conservation Aspects of Recreation." There is a difference between them. One deals with land as a total entity with a future, the other with satisfying people in the present. For example, conservationists have pleaded for the zoning of the south and southeast arms of Yellowstone Lake in order to protect waterbirds and the pelican rookery from speedboat disturbance. But under pressure from motorboat manufacturers and legislators, the Department of Interior continues to permit boats and is proud of its large new marina "to better accommodate the ever-increasing number of boaters," though I question why boats are allowed at all on a lake designated for preservation rather than mass use.

At this point, may I interject an observation on events of the very recent past [particularly the forced resignation of Conrad L. Wirth, director of the National Park Service] and statements by officials of the Interior Department on the future of the national park system? From here on, to quote one newspaper report, efforts will be intensified to "open up more parkland wilderness for development as recreation areas to serve a growing population."

Is this the proper role of the national parks? I would much rather see a rededication of the national park system to the principles upon which it was founded, as a composition of priceless vestiges of scenery and history, each one bearing the weight of overpowering national significance. I would much rather feel that the direction for the future provides greater, rather than less, protection and preservation of the national parks, free of political considerations, free of commercial pressures, and a reversal of the marked present trend to lower standards and to dilute the system with the addition of playgrounds called national recreation areas or some such name.

Of course, there should be provision for recreation developments. This is what we look to the Bureau of Outdoor Recreation to accomplish by giving guidance and leadership to the states and private landowners. But I find it frightening to hear the rising demands by commercial interests, and in turn by their legislative representatives, that the federal government pay the bill by establishing national parks in their regions and by altering the character of parks that now exist. One day last year I went out to see a proposed new area in company with a chamber of commerce man, who explained his complete endorsement in the following terms: "If we get a national park here, every roadmap in the country will be marked with a green dot. Vacationers will come in droves, and our section will earn millions in income." Where will it end? When green dots sprawl across the map and they are virtually meaningless?

The facts of population pressure and the shrinking land make the national parks all the more precious. They accent the challenge to protect the parks as sanctuaries of the American soil and spirit. I hope the Department of Interior will exercise boldness in facing this true challenge, rather than pursue a course of opening parkland wilderness and enlarging the national park system below its standards. Certainly the department deserves great credit for establishing the Bureau of Outdoor Recreation and giving it full support. Let the bureau inspire the states to do a job in the recreation field, which only few of them are now doing; but I hope the national park system may be reconsecrated to fulfill its fundamental, time-honored role.

There are some in Congress who take their conservation, as well as recreation, seriously. Recently Senator Gaylord Nelson of Wisconsin made what to me was an intensely impressive and important speech before a Washington audience, in which he warned that conservation of woods, lakes, streams, and other natural resources constitutes the most urgent and crucial domestic issue facing the nation today. But to find a report of his remarks in one of our fine Washington papers, the *Post,* I had to look to the bottom of the obituary page, where it lay buried below the deceased, like a very dead issue.

Someone tallied the total of syndicated Hollywood columns at ninety-nine and the total on conservation—that "most urgent and crucial issue"—at zero. There are a few newspaper people who write of the subject with knowledge and perception—John Oakes of New York, Roger Latham of Pittsburgh, Edward Meeman of Memphis; scarcely any magazine writers, outside the outdoor publications, since the passing of Bernard DeVoto and Richard Neuberger; hardly any authors of major books beyond Rachel Carson and Justice Douglas. And while you may disagree with what these people have to say, they have done their best to stir a public passion in conservation. The preference by the press as a whole for the glossy and glamorous is a sorry commentary on our times. Moreover, as long as it treats conservation—water, timber, wildlife, grazing, mining, recreation, the overall values of the shrinking land—as an incidental, now-and-then side issue, the public will remain in the dark. The truth is that if we cannot interest newspapers and magazines in giving thorough coverage to conservation, it becomes even more incumbent upon the public to assume the responsibility of informing itself.

In our quest for answers we look to the federal and state agencies dealing with natural resources. From some of them and from industries represented here as well, we are verily deluged with varied information, each assuring that its own brand of conservation is the very best. In a sense, this is good, for they provide us the opportunity to weigh their policies and actions and to judge for ourselves. But inevitably they lay the heaviest stress on the new facilities they are providing, whether dams, boat docks, campgrounds, ski lifts, or roads. What they are playing is a numbers game, and a dangerous numbers game, it appears to me, in which we the users are being used—and left enmeshed in confusion as to how they are managing our resources as a total entity.

So it is up to the users to go forth and observe, to perceive, question, challenge, and decide, to tell our congressmen what we want rather than have them advise on all they are doing for us. As campers, we should understand the why and wherefore of timber cutting around us in the woods, and whether it is good, bad, or indifferent to the land we are using. As fishermen, we should concern ourselves with steep-slope strip mining, which fills our streams with silt and kills fish with acid poison. As motorists, we cannot turn away our faces from the blight of billboards and other commercial intrusions that bring profit to a few but scar the countryside that belongs to us all. And curiously, some of the very congressmen who vote against billboard control are among the first to ask the federal government to build beautiful parkways in their states. All of us who look for clear water must be concerned with the evils of pollution and trace them to their sources. The hunter cannot avoid concern with management and protection of wildlife, and neither can the lover of animals and birds for their own sake, that maligned soul who is derided in the crush to mass recreation as an "esthete" and "emotionalist" but who stands out for his willingness to share a portion of the outdoors with wild creatures.

The recreational user must find the means to make his voice heard, as a responsible citizen deeply concerned with the well-being of the country as well as his own immediate welfare. We recently had the experience of Assateague Island, in nearby Maryland, where the landowners were virtually the only group in the state that expressed itself. My impression is that the state failed in its obligation to protect its own resources and that the agreement to establish a national seashore came only after it was reckoned that a flood of tourists would bring in new money. Conservation, unfortunately, was not an overriding consideration, and the Maryland public generally did not bestir itself. Now there is underway a study of the administration of the North Cascades being undertaken by two departments of government. I wonder whether we are going to let these two departments, with their pulls and tugs, resolve the disposition of the North Cascades. The public ought to know, to understand, to weigh what we may be getting, what we may be losing, what the implications are for the future in terms of all of our needs for natural resources in the decades ahead. Another circumstance involves the Allagash Waterway in Maine, where a tremendous area has been proposed for transfer from private hands to public hands. Whether or not this is sound ought not to be decided solely in the state of Maine, or on the basis of filling a recreational need alone, but on the basis of providing the wisest use of the land and the best means to conserve it. If the American public were not groping in the dark for an understanding of conservation, both these questions would be answered on a far higher plane. And I hope that all the issues and implications will be far more broadly aired, so that the private citizen may have the opportunity to express his viewpoint.

Every conscientious recreational user should align himself with a force of expression, whatever it may be, whether the Appalachian Mountain Club in New England, the Mountaineers in Seattle, the Sierra Club, the Audubon Society, the Izaak Walton League, or the others like them. They are yet today small voices in the wilderness; they must grow and give the public a sense of participation in the effort to safeguard and enhance the conservation aspects of recreation. Even so, when they are heard, they are heard well, as when the Audubon Society won its battle to protect the golden eagle from extinction, and when the sportsmen groups in Pennsylvania quit listening to the politicians and insisted the politicians listen to them on the need for strict regulation of surface mining, and when the Izaak Walton League in Oregon spoke out against political interference in the administration of grazing lands. You may not like what these groups have to say all the time, but they are the rallying points for the people, and I have confidence that given the light, we shall find the way, that if right is on your side in the conflict for land use, we shall recognize it.

As a member of the American Forestry Association, I am proud of its sponsorship of this congress, a forum presenting all the viewpoints on conservation. The association has tradition, a background of experience and wisdom; it has no ax to grind for any special cause except the well-being of the nation. But this meeting should be only a starting point, from which to tell the story and needs of conservation to a far broader audience than its own membership and encourage the public at large to exercise interest and responsibility as private citizens.

Speaking as a recreational user, I have learned that I must look beyond my own campfire, and beyond my small tomorrow. What I have tried to do here is give voice to idealism and integrity, a feeling for conservation without which outdoor recreation will become meaningless. Around me in the woods are many others, whose fires of understanding and action are waiting to be lit.

Talking to the Rangers

At Mather Training Center of the National Park Service,
Harpers Ferry, West Virginia,
March 28, 1968.

Following publication of *Whose Woods These Are* in 1962, Doubleday asked if I would consider writing a book about the Great Smoky Mountains of North Carolina and Tennessee. I had visited that region and written about it, though scarcely considered myself an expert. Nor was I mountain-bred. But if that didn't matter to the publisher, I would not let it matter to me. My editor, Samuel Vaughan, sent me an encouraging letter, advising that if I was able to write "up," then I might produce what he called "a small classic."

That is about how it turned out. Since the book first appeared in 1966, *Strangers in High Places* has been through various printings and editions and still sells respectably. Whenever I visit the region somebody comes to me with a copy of the original edition (price, $7.95 hardback), handed down by a parent or grandparent, and asks for an autograph; or, as one old fellow said, "I am pleased to meet you but I was sure you was dead."

I think the success of that book came not from the coverage of history, geography, geology, or anything purely factual, but from the portrayal of people living in what was long perceived as a closed mountaineer community. I've been asked how I, as an outsider, was able to make it with the natives. The truth is I can't remember a single unpleasant incident or ever once being turned away. I do remember the pleasures of listening by the hour to the old bear hunter and the mountain musician and the banker in Bryson City and the mountain doctor and the old Cherokee chief and the shaman and the loggers, rangers, and naturalists, professional and otherwise, now regretting only that I failed in those days to travel with a tape recorder. I mean that the voices I heard are gone forever, and that reporting them on paper to be read comes nowhere near as valuable as recording them to be heard.

I did have a problem connecting with the moonshiner in the book. My efforts seemed hopeless, especially after I had been canvassing the backwoods with the revenuer. Luckily, Ted Seeley, the district forest ranger at Brevard, whom I knew and turned to for help, was a connoisseur of mountain whiskey and arranged for me to meet a fellow who made it for a living. It was in the backwoods near Rosman, North Carolina, the very area I had covered with the revenuer. But then, during the fateful interview that lasted all afternoon, I had to listen to my host say, "Have another," and the only allowable answer was, "It sure is good."

Of all the people I met I was influenced most deeply and lastingly by Harvey Broome. He was a Knoxville lawyer, president of the Wilderness Society (one of its founders, in fact) and the leader in the political battle then underway to protect the wilderness of the Great Smoky Mountains. He inspired me to fulfill Sam Vaughan's challenge to write "up." Harvey died in 1968, but many years later, in 2001, when the University of Tennessee Press republished his personal journal titled *Out Under the Sky of the Great Smokies,* I was privileged to contribute the foreword. "Here we find Harvey, the wilderness apostle, on his home turf," I wrote. "He reveals himself exactly as I knew and loved him: a gentle spirit, sensitive to the needs of nature and humankind, always with tolerance and good humor."

The mid-1960s were a time of disillusionment and violence in America, rather than of tolerance. The country was deeply divided over the war in Vietnam, a military misadventure which to me legitimized mass killing and killing without cause. I mention below in the first paragraph of my talk at Harpers Ferry that in the week ahead I would be at Yale University. And the very day I arrived to speak at Yale, Martin Luther King was assassinated. Black ghettos erupted in fury and cities burned. I could not foresee the tragedy and its fallout, but I think I tried in this statement to deal with human issues and concerns and with the responsibilities of resource managers beyond resources at hand.

I was invited several times in these years to speak at the Mather Training Center (named for Stephen T. Mather, the first director of the National Park Service) at Harpers Ferry, West Virginia. I was already identified as a critic, but Raymond Nelson, the director of the training center, was an experienced hand committed to park principles who wanted to stimulate personnel with challenge rather than superficial stroking.

While I met at these training sessions with professional personnel at various levels in the ranks, I found almost all very cautious about speaking out, reflecting the power of the peer group in a structured institution. I told them they ought to be proud of a young ranger in the field named Edward Abbey, who had lately published a provocative book called *Desert Solitaire,* and emulate his honesty of expression. In due course Abbey left the Park Service. Ray Nelson was transferred to obscurity in the Washington office and retired to live the simple life in Maine.

■   ■   ■   ■

This is my second of three meetings in three weeks to discuss the public role in land-use policy. Last week I met with the forest supervisors of the Eastern Region of the Forest Service at Milwaukee, and next week I will be at the School of Forestry at Yale University. The audiences are different, but in one sense the same: concerned with land management and use, today and in the future. Since I endeavor to identify with thoughtful public interest in the land, I interpret the invitations to come before these groups as a desire on your part to know what the thoughtful public wants and needs, and what the relationship of the professional land manager, genuinely committed to public service, and this public must be.

My message to you is basically the same, and it is not even original with me. I borrowed it from Stephen C. Clark, the late benefactor of the New York State Historical Association, which runs the Farmers Museum and the great folk art collection at Cooperstown, New York. He insisted on authenticity, simplicity, and quality. His directive was, "If you offer people excellence they will find it out and respond to it."

To state it another way, let me borrow from Gifford Pinchot, who wrote that "the rightful use and purpose of our natural resources is to make all the people strong and well, able and wise, well taught, well fed, well clothed, well housed, full of knowledge and initiative, with equal opportunity for all and special privilege for none."

That means we are involved in a great social movement, serving all classes of people, appreciating their true needs, with faith in the intellectual capacity of people, even of the lowest classes—perhaps more of the lowest classes than any, for they are closest to the realities of life. In short, the day of idealism is not done; although idealism is hard pressed to survive by technology, shortsightedness, self-interest, and political venality.

Idealism is sustained in the land. It must be husbanded by managers of the land who have imagination, boldness, passion, and devotion to public service and the wisdom to reach out and touch the thoughtful public for ideas, cooperation, and support. These qualities are not easy to come by in this time of sameness, when it is easier to play the game for safety and security, but without them the public will be lost, and so will the land in the long run.

It is very important, I think, to connect philosophically natural resource management to broader social concerns, but I see little evidence of that being done.

Recently I read where one of postwar Germany's leading poets and writers, Hans Enzensberger, had resigned from a position at Wesleyan University in Connecticut in protest against U.S. policies. "The ruling class of the United States, and the government which carries out its policies, is the most dangerous collection of men in the world," he wrote in his letter of resignation, plainly referring to the misadventure in Vietnam. "Many Americans are disturbed about the situation in their country. In many respects the situation reminds me of that of my own country in the middle of the 1930s."

Soon after, I traveled to Georgia and Texas. In Atlanta I read about the appearance of Dr. Timothy Leary, the apostle of LSD, at Georgia Tech. He spoke to overflow crowds of young people; so many could not get into the auditorium for his scheduled lecture that a second one was held later in the evening, and it, too, was standing room only. He told his audiences:

> The sex revolution is part of young people's fight for freedom, and LSD is a symbol of the revolution. Kids today are not drinking like their parents were at their age. They are smoking pot and taking LSD. The fact that dope is illegal is great because that makes it an act of passive rebellion against the war and the lies of a robot, assembly-line society.
>
> The United States as a political entity is through. My job is to make it as gentle as possible, avoiding wars by men like General Westmoreland [William Westmoreland, U.S. commander in Vietnam].

Then I went to San Antonio, where it became clear that not everybody feels that way. Almost like a retort to Dr. Leary, the *San Antonio*

Light and News reported to its readers, "Radio KUKA, Spanish language station, now goes with red, white and blue, patriotic programs featuring American hymns, explaining the Constitution, stuff like that. It's real good. Let's just hope the whole-hog liberals don't start labeling the KUKA people 'bilingual Birchers.' To these libs, the only items worth discussing nowadays are what's bad about the country."

Whole-hog or half-a-hog, liberal or conservative, I like to think what's really good for the country is manifest in preservation of parks and historic places. Before the world, the conservation of our resources and measures like the Wilderness Act of 1964 demonstrate that ours is an enlightened nation. Preservation of parks and history embodies an affirmation of America's faith in its destiny, and of man's belief in himself. Even Dr. Leary would agree to that.

This is why what you do, what your goals may be, both personally and in terms of your organization, the National Park Service, are important in the broadest dimensions.

You are more under more pressure now than were your predecessors who ran the parks a generation or a decade ago. They had to have courage, to be sure, to face land exploitation and abuse of the parks. But it takes more courage today because the land is running out; the parks are prime for exploitation. And the people are flocking into the parks, not simply because they have leisure for pleasure, but because they are searching for their share of breathing room, wilderness, the better things about America, which are becoming increasingly compressed.

So you are being tested every day of your working lives. And it isn't always easy to pass the test.

Let me tell you of an incident in the course of my travels in the national parks. I met a young assistant ranger, a good looking, knowledgeable fellow, who was working under a friend of mine. "Why don't you stay here and learn this wonderful country, and then take over from my friend?" I asked him. "I will explain it to you," he replied. "You see, I am working toward my retirement"—he was a young man less than thirty years of age—"and if I can advance, then my retirement earnings will go up. Therefore, as soon as I get an opportunity for a better-paying job in another park I will take it."

I sympathize with the desire for advancement, and the need to provide for one's growing family; but I suspect that young ranger will not come through, whatever position he may reach, high or low, for he will spend his career cultivating his retirement by playing it safe.

When I was doing research on my book about the national forests a few years ago, I spent some time with a ranger in Montana, John Hall by name, who subsequently became a good friend and is now on the staff of the Wilderness Society.

"Oh, I guess I could quit the Forest Service and earn more in private industry or on some other line, but I doubt I would be any more content," he said. "My family may never have everything in life, but I can provide for them on a ranger's pay. I develop a campground, build a trail or conduct a timber sale, and feel I've done something for my country and the coming generation. Money can't buy that feeling. After all, there is a psychic income in knowing you will leave the land in better condition than you found it." It is the psychic income, the fulfillment of one's idealism, the vision of the people's tomorrow, that should be implicit in the goals of all working for the National Park Service. Occasionally, I know, some who stand up for what they believe get hurt. I have friends in government agencies, including yours, who have lost their chance for advancement by sticking to principle when political expediency dictated otherwise. But the pursuit of principle is its own reward.

Let me give you examples, from your own ranks, of what I mean:

A naturalist in the Everglades, a young woman named Gale Zimmer, whom I have never met, recently wrote an article in your internal publication *Trends* titled "Concerning Dangers in National Parks," inspired by the grizzly incidents in Glacier. [In 1967 two women campers were clawed to death in their tents, after which two bears assumed to be involved were destroyed by park personnel.] She wrote this while the Washington office was occupied in counting the pro letters and the con letters and justifying to all the murder of the grizzly bears. The management of Glacier, the national park involved, and of the regional office appear to me to have held to their own close counsel. But Gale wrote thusly, and bravely:

> Maybe danger belongs in a National Park. I think its being there is a part of what we mean by a "wilderness experience," a "national park experience." National Parks are not cozy roadside tourist attractions, designed to satisfy the curiosity of mankind in padded comfort. They are slices of the natural world and thus they should be. In the natural world there is "danger."
>
> People should know what a national park is—and isn't—before they commit themselves to spend their vacation or weekend there. They should know if it's

going to be rough and primitive and if they want the rough and primitive and natural, fine. I think we have an obligation to inform people—honestly. But I think we betray the ideal behind the whole National Park System if we try to plane down all the rough spots, shoot all the touchy animals, fence off all the cliffs and offer visitors "a National Park scene in the safe comfort of your own living room." With Thoreau I'd like to know an entire heaven and an entire earth, and I think basically our natural National Parks should offer an entire heaven and an entire earth.

Thoreau would be proud. We all should be proud. That is Gale's reward.

Next we have Edward Abbey, ranger at Arches National Monument, whom I encountered in a column of the March 1968 issue of the *Jackson Hole Villager,* as follows:

Mr. Abbey doesn't like automobiles. He has it in for the internal combustion engine and has misgivings about the manners and morals of those who have become dependent on its use. He doesn't like bumper-to-bumper traffic and has reservations about park administrators who want to build more roads. Mr. Abbey could lose his job.

Abbey offers a solution that is about as palatable as the Russian cure for the common cold: "No more automobiles in national parks," he writes. "Let the people walk. Or ride horses, bicycles, mules, wild pigs—anything—but keep the automobiles and the motorcycles and all their motorized relatives out."

Very few are going to buy Abbey's argument. After all, we're committed to cars, Hondas, snowmobiles, fireworks stands, another length of runway, four-lane highways, spin-fishing, continental breakfasts, bridges over rivers where the ducks now sit, more parking for the post office, TV in every room. We can't go along with you, Abbey, and we can't do what you ask, but we need men like you on the other side to reproach us lest we, in some frightful spasm of energy, throw all the sand out of our playbox.

You should all be proud of Edward Abbey and emulate his honesty of expression. I for one only wish that we had him, or one like him, in the leadership of the Great Smoky Mountains National Park during our period of struggle to save that park from the blessings of roadmanship.

Then we have Kenny Dale of the park practice program. I do not know what took him to look at Overton Park in Memphis, where decent members of the community have been fighting a valiant battle to save their inner city parkland from the bulldozer against an alliance of commercial businessmen, real estate speculators, highway engineers, and a monopolistic, venal press. The Citizens to Preserve Overton Park interested me in their case and asked if perchance there was some agency in Washington to help them since federal money is involved. But almost everyone in Washington said simply, "But we cannot interfere unless the political climate is favorable and a local political body asks us to." In any event, Dale went to Memphis and gave this report while there:

> The transcendent value of Overton Park for park purposes should be obvious to anyone. Less perhaps are the irreplaceable values inherent in the wooded portion of the park. My observations lead me to believe that the potential recreation and nature education opportunities available in this portion of Overton Park have not begun to be realized. . . .
>
> In my opinion, it is a mistake to use this park for highway purposes, and I am sorry that I can do nothing to prevent it. . . . It is not likely that a city administration that permits the destruction of such values would make much effort to provide them again.

His statement came at a critical time; it gave heart to the park defenders and pause to the park destroyers. It may still prove the factor that turned the tide. [The Overton Park case ultimately was debated in Congress and before the United States Supreme Court, which ruled in favor of the Citizens to Preserve. That park remains an asset to Memphis.]

How many people in the National Park Service would issue the same kind of statement? How many are ready to commit themselves openly, like Edward Abbey, on the question of automobiles in National Parks? How many are willing to write about our true relationship with wildlife as Gale Zimmer did? I came here to Harpers Ferry to find out; I came here also to say that you must be "turned on," with a sense of mission, setting your sights on excellence, the highest quality, as your gift to the people, because they deserve it, rather than to accept and transmit mediocrity and to compromise your idealism. You will find more allies and supporters—especially those worth having—from strength of conviction instead of weakness and the so-called consensus.

As one of your allies, I believe that we must begin together to bring light into the darkness of slums. It does our social movement no good at all to save the large wildlands if we fail to bring air, space, and a community of growing things, other than rats, to the poor people in the ghetto prisons where they live. For one thing, if people do not find space to camp and fish and see trees close to home then they are driven into the woodlands and wilderness. On this basis I have criticized my friends in the Department of Agriculture for their willingness to surrender forests and farms around the edge of cities and their promotion of industry in rural America.

But even more important, I recall that two years ago I went to the opening of the new baseball stadium in Atlanta, a city of new affluence. Before the game began there a lavish party was given by the ballpark management, with plenty of booze and food. However, I remember most the pickets in front of the stadium. They were eight or nine years old, dark of skin, and carrying placards reading "Where are the parks we were promised?" These are our most important clients, the human resources, asking to be conserved, and thoroughly deserving.

I could go on and expand on these themes, but I believe we are to discuss them on the basis of this paper. There is one other thought I would like to set down: your cooperation with citizen groups—better yet, your participation in citizen groups. Alas, I see very few National Park Service people (or Forest Service people either), in uniform or out of uniform, at affairs of the Wilderness Society, Sierra Club, Appalachian Trail Conference, Audubon Society, Izaak Walton League, Northern Virginia Conservation Council, or like groups.

Many of you are cut off from the mainstream of people who care and therefore weakened in your efforts to manage the lands to the best of your ability. When we hiked in the Great Smokies one year ago last October, in order to demonstrate our devotion to wilderness, only one person from the National Park Service was with us—and he was on official duty, riding a horse.

    I would much prefer to be constructive than critical simply for the sake of criticism. I went to interview Robert Frost during the last year of his life. Among many wise things he said that day was, "Americans are, by their heritage, splendid risk takers, like good gamblers." In the constructive spirit, I warn you that it is disastrous to sacrifice standards and to lower your sights in order to meet the lowest common denominator. Now is the time for risk taking in the people's behalf.

Insects and Weeds Need Friends

At the annual meeting of the
Society of American Foresters,
Las Vegas, Nevada, October 14, 1970.

In December 1999 I joined the protest against the World Trade Organization in Seattle. Surrounded there by a bubbling, buoyant, massive throng of young people, I thought back to another historic event, the first Earth Day, almost thirty years before. Both these actions and other experiences over the years have shown me that action for the good is what many young people crave. They want to give rather than be given to, to serve rather than be served, and to help shape a democratic society on principles of equality and justice.

I loved Earth Day 1970. It came on the crest of rising environmental awareness. Thanks to Earth Day, the environment was recognized as a valid and important issue. It became acceptable to teach about it in classrooms, to write about it in newspapers and magazines, to talk about it in board rooms, to campaign for it in politics. Gaylord Nelson, who conceived the idea of Earth Day while serving in the U.S. Senate, declared, "Now is not the time to assess guilt, but rather to assign responsibility," allowing everybody to claim a stake in it.

I went around and made speeches on four college campuses—first at Earlham in Indiana, then at Ohio State, Denison, and Ohio University—not on the same day but in the week of the Environmental Teach-In. The crowds were large, responsive, and upbeat. I made other Earth Day speeches here and there in later years and enjoyed being with the students, feeling this was *their* day and *their* time in life to be heard. Many of them have gone into careers in other fields, but a few, I'm sure, have stayed with it, and all society has benefited.

I loved the WTO protest, too, although the sweetness and optimism of Earth Day were gone and the young looked younger to me than they did thirty years earlier. Pundits of the mainstream media tried to trivialize the "Battle in Seattle," capitalizing on isolated incidents of petty violence. But the coalition called People for Fair Trade brought together forty-thousand-plus protesters and observers from around the world representing movements of labor, the environment, animal rights, peace, farming, religion, and children's rights. They merged in Seattle in a common cause, demonstrating concern about corporate power in globalization as manifest in the WTO. Although established by governments, the WTO operates with greater authority than governments, shredding the ability of individual nations to enact and enforce laws designed to protect the environment, workers, children, animals, human health, and human safety. That system was devised behind closed doors; the protests in Seattle and others that followed elsewhere forced it open to public scrutiny.

In 1970, I was writing columns regularly in *Field & Stream* and *American Forests,* so that foresters and other resource professionals knew me. Forestry, which once had been the leading edge of conservation, now was at the tail, aligned with commodity production rather than protection and preservation. Still, somehow, I was invited to speak my piece before the professional society.

I was asked to present the environmental view of pesticides. Nobody has done that better, in any period of time, than Rachel Carson. Her book *Silent Spring,* published in 1962, was a landmark of literature, science, and ethics. Personally, she was modest and retiring, concentrating all her energy on her work. She was not a joiner but came on the board of directors of Defenders of Wildlife (of which I was already a member), but barely served before she died in 1964 at the age of fifty-six. In an edition of *Silent Spring* published after Carson's death, her editor and friend, Paul Brooks, wrote, "Rachel Carson was a realistic, well-trained scientist who possessed the insight and sensitivity of a poet. She had an emotional response to nature for which she did not apologize. The more she learned, the greater grew what she termed 'the sense of wonder.' So she succeeded in making a book about death a celebration of life."

Despite seeming victories in public awareness and government action, even after all these years the use of chemical poisons remains a critical issue in society. DDT is banned in the United States, but other poisons are used here in its place, while DDT is applied elsewhere in the world and products contaminated with it are shipped to us for consumption. Now pesticides are registered, as required by law, but it doesn't follow that they are safe or any less damaging.

In the year 2000 the U.S. Geological Survey reported finding pesticide residues in 95 percent of water samples drawn during the last decade from twenty river basins across the nation. In December 2000, forty years after the insecticide diazinon was registered for use, the Environmental Protection Agency finally announced an agreement with manufacturers to slowly decrease its availability. EPA called it "one of the leading causes of acute insecticide poisoning for humans and wildlife," but the best it could do was to phase out sales of diazinon for use in lawn, turf, and garden products and for less than half of all agricultural uses. Besides, EPA's chemical-by-chemical approach means restrictions on one hazardous chemical simply increase the use of others.

The chemical industry did its best through fair means and foul to discredit Carson and has never let up on anyone who dares to take it on. Sandra Steingraber, a scientist drawing from her own experience with cancer, wrote *Living Downstream: An Ecologist Looks at Cancer and the Environment,* published in 1997. After a string of favorable reviews, she was startled to find her book panned in the esteemed *New England Journal of Medicine*—until she discovered that the reviewer, Jerry H. Berke, was not only a physician but also medical director of W. R. Grace and Company, a chemical company cited in her case histories. Even worse, the *Journal* editors were aware of the connection.

In the same year, 1997, the Center for Public Integrity in Washington, D.C., published a report on *How the Chemical Industry Manipulates Science, Bends the Law and Endangers Your Health.* It showed how top EPA officials ended up working for chemical companies, their trade associations, and lobbying firms, and how a former chairman of the Consumer Products Safety Commission, John Byington, helped the chemical industry launch a campaign to neutralize his old agency. The Public Integrity report documented free trips given by chemical companies to congressmen and EPA officials; it showed these firms employing nearly 90 percent of the nation's weed scientists, and the few independent researchers relying heavily on grants from pesticide manufacturers.

These manufacturers, their allies, and apologists tell us that intensive industrial agriculture is really like a public service—absolutely necessary to grow enough food to feed the world's hungry. They don't say their methods actually reduce yields, poison and erode soils, degrade water quality, and around the world force indigenous peoples to shift from sustainable diversity into acceptance of the mislabeled "Green Revolution," an alien industrial system that does not work.

On the other hand, a lot of people have learned of the damages caused by toxic chemicals—including cancer, birth defects, nerve damage, and brain damage—and are asking for better, nonchemical ways of dealing with what we call "pests." The growth in organic farming and organic food consumption represents a significant response. Environmental and public health organizations are helping the public to understand what product labels mean and do not mean. These groups and many labor unions are working to restrict the use of genetically engineered crops and to replace pesticide use with sustainable methods of growing food and fiber.

The ultimate answer, if you ask me, will come when the idealism of Earth Day and the WTO protest transcends corporate power and influence. A tough road, but nobody said life was easy.

■　■　■　■

The conservationist's essential cause is the protection and perpetuation of life, not simply his or her own life in its brief, transient human form, but the cause of life everywhere, in all its forms. His or her concern is for the interwoven life-environment of land, water, and sky, embracing the material, tangible essence and the intangible cosmic, or spiritual, unity.

The idea may appear misplaced in this hectic age when death is commonplace—death on the battlefields, in the cities, on the highways, and in the conquest of nature. But life-love is an old, old idea. It goes back to the pre-Confucian Chinese who described a society wherein "not only a man's family is his family but all men are his family and all the earth's children his children." And to the psalmist who spoke of God: "For every beast of the forest is mine, and the cattle upon a thousand hills. I know all the fowls of the mountains: and the wild beasts of the field are mine."

[I might have quoted a passage from Isaac Bashevis Singer's classic novel, *The Slave,* published in 1962, the same year as *Silent Spring.* Singer's hero, Jacob, was working in the fields late one day, reciting biblical passages to himself:

> [As the scythes moved, field mice ran from their
> blades, but other creatures remained in the harvested
> fields: grasshoppers, ladybugs, beetles, flyers and crawlers,

every variety of insect, and each with its own particular structure. Surely some Hand had created all this. Some Eye was watching over it. From the mountains came grasshoppers and birds that spoke with human voices, and the peasants killed them with their shovels. Their efforts were to no avail, since the more they killed, the more gathered. Jacob was reminded of the plague of locusts that God had visited upon the Egyptians. He himself killed nothing. It was one thing to slaughter an animal according to the law and in such a way as to redeem its soul, another to step on and crush tiny creatures that sought no more than man did—merely to eat and multiply. At dusk when the fields were alive with toads, Jacob walked carefully so as not to tread on their exposed bodies.]

Of all the forest technologists I know, very few at most either grasp or accept this meaning of life-love. They consider themselves to be very practical people; they appreciate the "realities" of economics and politics. They either tolerate or deride the conservationist as an idealist or a pantheist or a do-gooder who doesn't understand how to get things done. But exactly what is it that must be done?

According to an official of the Weyerhaeuser Company, Royce Cornelius, speaking at a forum on forestry education, the obvious function forest resource management graduates will fulfill in industry is to keep the wood basket filled and at the same time to be cost-conscious and to return a profit. "Time is money in industrial forestry," Mr. Cornelius declares. "To be competitive, every forest acre must be working to its capacity in producing wood fiber."

That may have been acceptable practice for yesterday, but it must be modified for tomorrow. If the United States is to endure and prosper as a nation, all land policy must henceforth be based on the foundation of environmental respect. The goals must be to provide a continuing supply of clean air, clear water, stable soil, natural beauty, and open space, including wilderness. The only alternative will be the continuing decline in natural resources, national morale, and ultimate failure. Thus we need to break with the past when land values were determined in narrow economic terms. We need to shape a new responsibility and respect for the total land resource.

This resource includes insects and weeds, the notorious pests, the unwanted in the practical world of people who get things done. It's amazing how little we understand or appreciate these life-forms. Many insects we destroy indiscriminately are regarded in more frugal countries to be good food. Not all insects are dirty either; some, in fact, are cleaner than the residues of insect killers that farmers spread on their vegetables. By the same token, many plants we shun as "weeds" yield wholesome, palatable, and edible fruits, nuts, leaves, stems, roots, and seeds to those who care.

Without weeds, erosion and flooding would be rampant. George Washington Carver found a weed was something good, whose usefulness man had not been smart enough to discover. The potato was once thought of as useless, and the tomato as poisonous. When a soil loses fertility we pour on fertilizer, or at least alter its tame flora and fauna, without considering that its wild flora and fauna, which built the soil to begin with, may likewise be important to its maintenance. Years ago it was discovered that good tobacco crops depend, for some unknown reason, on the preconditioning of the soil by wild ragweed. The same may apply on a broader front for all we know. Moreover, I think of the beautiful weedy hillsides and swamps, the views of black-eyed Susans, white daisies, arbutus, and trillium, the wonder in the field of blooming dandelions, a weed if ever there was one.

Every effort is being made to eradicate such creatures from the agricultural scene. Farming has moved from a family way-of-life to a factory-in-the-field, based on large machinery, fertilizers, and intensive monoculture, the source of profits to investors and industrial syndicates far from the land. Monoculture on the one-crop farm, or one-crop anything, favors the irruption of pests, as does any land use which greatly simplifies the ecosystem. Insect pests that result from the monoculture are becoming more of a problem every year. More toxic materials must consequently be used to cope with blights and plagues. The nitrogen cycle is broken, and soils lose their nitrogen-fixing bacteria. The land mechanism becomes like the dope addict who needs a shot to function at all. Industrialization of agriculture is speeding up the biological functions of the life-community to the point where it cannot maintain a healthy environment for the growing plant.

All of this is done with sanction and encouragement from the Department of Agriculture, which is certainly no model of ecological responsibility. Despite its continuous insistence that chemical poisons are necessary to food production, individuals and local organizations all over the country are demonstrating this is not the case. One example is the 107-acre farm in Miami County, Kansas, managed by Dr. E. Raymond Hall, the noted biologist of the University of Kansas, growing a full quota of corn, milo, soybeans, and wheat, untouched by poisons and with weeds controlled by the simple device called cultivation. The system produces more bushels per acre and more total bushels than the farms around it. "Furthermore," advises Dr. Hall, "under this system the fish in the creek grow big and do not die prematurely because of residues of pesticides that already have stilled the spring voices of five bird species in the woodland on and around the land that I manage."

As for our forests, plantations are now producing one-log or two-log trees on soil that originally grew three-log and four-log trees. The difference lies not in the tree but in the microflora of the soil that take far more years to restore than to destroy. An acre of healthy soil is densely populated with millions of insects and tiny animals, diligent architects and engineers sifting air and water, returning valuable nutrients to build the soil in support of higher life-forms and transferring energy along the food chain. Insects and animals benefit plants by carrying pollen. Wherever insects are abundant, other insects come to eat them. An abundance of insects produces an abundance of birds, a fact generously ignored or unknown in the widespread dispensation of chemical poisons.

The conservationist recognizes the mechanism in the "pyramid of numbers," that is, a very great number of very tiny organisms at the base, supporting a slightly smaller number of still larger ones, and so forth, with each level depending on the one below it. The tragedy, to paraphrase Dr. Howard Ensign Evans of Harvard, is that "practical" members of the human society at the top of the pyramid have difficulties understanding the creatures on the bottom, though without them the whole edifice would come tumbling down.

"Forest entomologists have directed a major portion of their time and talents to finding ways to put the insects down." So declared the chief of the Forest Insect Research Branch of the Forest Service, William Waters, at the meeting of the Society of American Foresters in 1969, as reported in the *Journal of Forestry*. He expressed his concern over both actual and what he called "threatened" losses from insects and lamented criticism of large scale aerial spraying of DDT and other hard pesticides.

He declared, "Most of us are familiar with successful uses of DDT and other synthetic chemicals in controlling a long list of insects extending from spruce budworm and gypsy moth to white grubs and other pests in nursery beds. In fact, there are very few forest insect pests for which no chemical control method has shown at least partial success."

Is that actually the case? The Forest Service, other federal agencies, and private landowners have sprayed millions of acres, with the inevitable result. Poisoning the ecosystem removes our allies (birds and predacious insects). Since prey build up faster than predator, it becomes necessary to spray again and again. Insects that were never any bother are released from predation, become pests, and must be controlled. Despite Mr. Waters's expression of pride and confidence, our forests were beset in the last year with many new varieties of insect pests. The 1970 recommendations in *Agricultural Handbook No. 331* lists thirty-four new insects not mentioned in the 1969 recommendations, and of those thirty-four new threats to our forests, the Forest Service recommends chlorinated hydrocarbons such as DDT as the controlling agent for twenty-seven.

But pests plainly have developed resistance to chemicals, while costs of pest control have increased and pesticides pollute the atmosphere. Pest organisms ordinarily have large populations—that, indeed, is why they are pests. "The large size of their population," writes Dr. Paul Ehrlich, of Stanford, "makes them much more likely to have the kind of reserve genetic variability which leads most easily to the development of resistant strains." And therefore:

> Extermination of a pest by use of synthetic pesticides is unlikely—and in fact, virtually unknown. The usual picture is one in which the pesticide decimates the natural enemy of the pest, while the pest develops resistant strains. These problems of pesticides are just one example of a trend, which has accompanied the expansion of the human population. The more we have manipulated our environment, the more we have been required to manipulate it. The earth has come largely under the control of a culture which sees man's proper role as dominating nature, rather than living in harmony with it. It is a culture which equates "growth" and "progress" and considers both as self-evidently desirable.

Now we may be facing the worst on our public lands. Either through the National Timber Supply Bill before Congress, the dominant use principle of the Public Land Law Review Commission, or the president's own directive on timber, receipts from timber sales are apt to be earmarked directly for single-use management. Under this circumstance, wildlife habitat will be sacrificed to speed timber growth. Food plants will be poisoned to remove them from competition with the so-called desirable, or most profitable, timber species. Game cover will be eliminated. If thickets are of inferior species they will get the "release" treatment. If thickets happen to be of a commercial timber species, they will be thinned to an even spacing and pruned. This would make easy shooting, but with food and cover eliminated there won't be much game to shoot at for long. This is called intensive, high-yield, dynamic, multiple-use management.

Item 6 of section 6 of the Timber Supply bill provides for using fertilizer to speed timber growth. The "most efficient" way to apply fertilizer to a forest is by broadcasting, probably from a converted bomber. Much of it will wash into streams where it will promote the growth of algae. Thus while the president has asked for a program for clean waters, Congress will consider authorization to foul up what little clean water we have left. [The National Timber Supply bill was roundly defeated, but timber-forestry advocates have never quit trying to define wood production as the primary use of public forests.

[Regarding the Public Land Law Review Commission mentioned here, William K. Wyant provides a brief, succinct summary in *Westward in Eden* (Berkeley and Los Angeles: University of California Press, 1982), 127–28:

> [Of the four studies of laws and public land policies made since the late 1870s, by far the most comprehensive and in many respects the most intelligent and useful was the last—the massive survey completed in June 1970 by the Public Land Law Review Commission under chairmanship of Representative Wayne Aspinall of Colorado. The Aspinall survey took about five years and cost more than $7 million. It published its findings in a bland and somewhat unctuous document called *One Third of the Nation's Land,* a staff product of the late twentieth century. The commission was weighted heavily on the side of mining, timber and other commercial uses.

[Despite his criticism, Wyant said the commission came up with the right answer "on the one big, burning, controversial issue"—by recommending the federal government retain in public ownership virtually all of the public domain lands administered by the Bureau of Land Management.]

Chemical poison, of course, has been implicit in forest technology for some time. In fact, I considered titling this talk "Can Forest Technology Survive Without Poison?" On July 25, 1955, the Department of Agriculture announced the Forest Service was spraying two and a quarter million acres of spruce-fir forest with DDT from planes at a cost of more than $2.5 million. The news release stated, "One pound of DDT mixed with one gallon of fuel oil is sprayed per acre. This light dosage will obtain a thorough kill of budworms, but the spray is not destructive to fish, birds, or wild or domestic animals."

The Forest Service had the chance to learn early and reverse course, for soon after there appeared in *Nature Magazine* an article by a scientifically trained observer, Vernon Bostick, describing the effects of such spraying in Carson National Forest of northern New Mexico. "The Forest Service claimed a 93 percent kill of budworms," he wrote. "Their success on other insects, except ants and bark beetles, seemed even higher. Every leaf of every plant on the forest floor seemed speckled with brown spots where the droplets of fuel had landed. . . . It is hard to imagine such vast acreage so completely devoid of insect life. I found all larvae killed, from the tiniest midges to huge, three-inch-long willow-fly nymphs. It was a thorough job."

Mr. Bostick raised these questions, which one may raise anew with reference to such chemical poisoning: With no insects, what would become of insectivorous birds? Adult birds could migrate, but how could they raise their broods? And what of game birds—particularly since grouse and turkey populations seem to correlate with the abundance of forest insects?

He also raised the question of whether spraying for spruce budworm constitutes good forest management. Spruce budworm is a native forest insect. It has been with us as long as our forests. We still have forests, and they are fairly productive. When conditions are favorable, insect populations tend to build up to epidemic proportions. This, of course, makes things favorable for their parasites and predators, and these also increase in number. The combination of reduced food supply—the direct result of overpopulation—and an increase in their enemies soon wipes out the epidemic and reduces the insect to normal numbers. Bostick declared, "The promiscuous use of DDT kills friend and foe alike. The budworms surviving the spraying (7 percent of an abnormally high population) are free to multiply, unhindered by their normal enemies. Heavier dosages will be needed. It would seem obvious that no agency can dump more than a thousand tons of DDT, and millions of gallons of fuel oil, from planes without creating havoc with the balance of nature."

Most of the modern insecticides and herbicides have been synthesized by chemists at the behest of chemical company managers and salesmen. The chemist knows how to mix poisons, I'm sure, but the salesman serves as diagnostician, therapist, and pill dispenser, with aid from collaborators in the agricultural establishment. He analyzes the pest problems, recommends the chemicals to be used, and promotes their sales through advertising and merchandising techniques. He doesn't have to demonstrate his professional qualifications, nor is he licensed. Yet hc meets ecological problems with deadly, disruptive chemicals.

EXAMPLE: The *Douglas (Ariz.) Dispatch* of February 13, 1970, reports that "weed control chemicals of value in this area will be discussed at a meeting at the Wilcox Veterans Club, Wednesday, February 18, 10 A.M. to noon. A buffet dinner will follow sponsored by the chemical companies. This is a joint effort between dealers in agricultural chemicals and the University of Arizona Cooperative Extension Service in Cochise County. Industry representatives will point out the merits of the chemicals they produce."

EXAMPLE: A lone gypsy moth is trapped in Shenandoah National Park, Virginia. No egg masses are found. No female of the species. No other evidence of infestation. An urgent meeting is convened at park headquarters. Those present include park personnel, representatives of the U.S. Department of Agriculture, Virginia Department of Agriculture, and the state agriculture college. There is no ecologist present. The meeting is not publicized, yet there are two outsiders, salesmen for Union Carbide Company, manufacturers of a chemical poison called Sevin. Both Massachusetts and Connecticut long ago gave up trying to eradicate the gypsy moth. New Jersey does not recommend spraying for anything less than five hundred egg masses per acre. Yet the Park Service acquiesces and surrenders its responsibility for management of a natural area. It then defends itself on the grounds that Sevin is "an approved non-persistent pesticide." It does not reveal that Sevin is patently nonselective, that it kills beneficial insects, threatens birds and other wildlife, and of itself can kill birds. Sevin is toxic to honeybees, the indispensable pollinators, with long-lived adverse effects upon the food chain. Although people should not be exposed to Sevin because it has caused a high incidence of birth defects and malformations in test animals, the Park Service approves spraying around the campgrounds of Shenandoah National Park.

The people are being led astray. They are not being protected or represented. One may wonder whether the purchasers of these poisons are shooting in the dark. In 1968 the General Accounting Office criticized the Pesticide Registration Division (PRD) of the Department of Agriculture [later moved to become part of the Environmental Protection Agency] for failure to initiate a single prosecution, despite evidence of repeated violations. In April 1969, the director of this division admitted at hearings in Madison, Wisconsin, that his agency registers pesticides mainly on the data supplied by chemical manufacturers without analytical checks of its own. On May 7 and May 24, 1969, the affairs of this agency were the subject of public hearings by a committee of Congress. I commend to your reading the report on *Deficiencies in Administration of Federal Insecticide, Fungicide, and Rodenticide Act,* issued by the House Committee on Government Operations, dated November 3, 1969. It tells a shocking story of abuses, laxity, and conflicts of interest involving PRD employees and pesticide manufacturers. Since passage of the act of 1947 until the time of the hearings, USDA had never removed a pesticide from the market over the objections of the manufacturer—except once, when Shell Chemical Company complained about a competitor's product, which was then suspended and removed from the market.

The House Committee found that HEW had objected to 1,633 proposed registrations or re-registrations in a five-year period. But neither HEW nor Agriculture could report to Congress how many of these products were registered, since PRD did not keep records or advise its sister agency. Safety warnings were found highly contradictory. For instance, products containing thallium sulfate—a toxic ant and rat poison—bore warnings that the poison should be kept away from children. However, directions for use called for the product to be placed on floors and in other places accessible to ants and rats—and obviously to children as well. Products containing lindane and DDVP bore warnings against contamination of food, but their labels indicated they were approved for use in restaurants under conditions certain to result in contamination of any exposed food. A fly and roach spray product cautioned the user, "Use in

well ventilated rooms or areas only." But directions for use began, "Close
all doors, windows and transoms." This reminds me of the following
printed cautions in the use of Sevin: "Keep out of reach of children and
animals. . . . Harmful if inhaled or swallowed. . . . Avoid contamination of
food, feed, water supplies, streams and ponds," followed by directions
for aerial spray.

The House subcommittee, completing its study in Washington, D.C.,
found PRD a dismal failure in its responsibility to protect the public from
hazardous pesticide products. Yet, as though to complete the hoax, the
garden editor of the *Washington Post* glorified the PRD and the use of
chemical poisons. On July 11, 1970, with the report easily available to
him, he wrote as follows:

> The USDA Pesticide Regulation [*sic*] Division
> employs a large number of well-equipped biologists,
> entomologists, pharmacologists, toxicologists, bacteri-
> ologists, and plant pathologists. One cannot imagine
> the amount of dedicated talent there is here. These
> people are experienced experts with access to every
> kind of detailed information about pesticides and with
> authority to require any additional information they
> may need from the registrant. They consult with the
> Public Health Service, with the Fish and Wildlife Service.
> They study the hundreds of pages that report the tests
> that have been made by the manufacturer before filing
> application for registration. They do testing themselves.
>
> These experts are not working for the chemical
> manufacturer. They are working for the people of the
> United States of America. If there were any way to make
> the system more fool proof, they would be the first to
> welcome it.

Distortion and suppression of the truth concerning chemical poisons are awesome and frightening. In the July 1969 issue of *Field &
Stream* I reported receiving word from concerned citizens in Tennessee
that Bowaters Southern Paper Corporation had spread poison grain over
a wide area for the benefit of a pulp-producing monoculture. I noted
that members of three hunting clubs had taken separate inspection
tours of Signal Mountain, outside Chattanooga, and all had declared it a
terrible tragedy for wildlife.

This evoked an anguished letter of protest from the public relations manager of Bowaters, accusing me of "misleading statements and fabrications." His viewpoint was supported by the Tennessee Fish and Game Department, U.S. Bureau of Sport Fisheries and Wildlife, and a wildlife biology professor at the University of Tennessee who had prescribed the poison, zinc phosphide. They all assured me in correspondence that environmental risks were slight and of a temporary nature, that all complaints had been carefully investigated but had proven groundless or insignificant. They insisted that only the targets—voles—were affected and there were no hazards to birds or game. But I was deeply disturbed that expressions from all of them on behalf of the economics of pulpwood production transcended concern for the biological and ecological resource.

I went on the ground and interviewed people in the mountains. They were upset. One man told me poison corn had been sprayed all over his property, which adjoins the Bowaters land. Another said, "I found poison in the water and was afraid to drink. People fear to hunt and eat game with poison in the system." I learned the report of Marzine Hudson, of the state office of the Bureau of Sport Fisheries and Wildlife, confirming the damage, was suppressed. I made the trip into the hills accompanied by a television camera crew from Chattanooga at the request of that station. The program has never been shown.

I consulted Victor Cahalane, E. Raymond Hall, and Clarence Cottam, biologists of international stature. All confirmed my fears. Cottam wrote, "If one can pick up any quantity of this and find it in piles, it is perfectly obvious that it hasn't been well put out. When poisoned grain is put out in that fashion it is perfectly obvious that it can't help but do a lot of damage to nontarget species. An aerial procedure of spreading poisoned bait, spray, or dust makes it almost a sure fact that nontarget species will be killed, and I think there is little doubt that a good many were killed in this instance."

Although Bowaters and its supporters insisted that zinc phosphide kills only voles, I found that every piece of current literature stresses this to be an intense, dangerous, and long-lasting poison. Warnings such as the following are plain and plentiful: "Zinc phosphide must be used with care, as it is toxic to all forms of animal life. It has poisoned humans as well as domestic and wild animals. . . . Be certain to remove and destroy all uneaten baits at the end of the poisoning period."

The day I arrived in Tennessee, Bowaters announced deferral of its second poison attack on the voles. This was wise. An ecologist would explain that total annihilation programs for rodents are rarely successful;

and those declines are followed by greater reproductive success. Instead of looking for panaceas in the bag of trick poisons, one might suggest careful ecological research and serious self-searching concerning forest management practices that may account for high vole populations. For example, I saw huge bulldozed piles of dead hardwood—an utterly wasted natural resource—serving as a clear source of protection for rodents.

I recently received a clipping from Tennessee, which showed I had been right all along. Dr. Ralph Dimmick of the University of Tennessee was quoted as reporting that dyed oat groats were less damaging to quail than cracked corn dyed green—which Bowaters had used. To quote the clipping, "He warned, however, that further testing should be conducted on the employment of zinc phosphide in widespread use for wild rodent control. 'Its high degree of toxicity and its non-specific nature obviously pose some hazard to all vertebrates that may encounter it.'" (This, of course, they all had repeatedly denied.)

"'Substantially more experimentation needs to be done to improve and/or determine its repellence to other species,' advised Dimmick. 'The search for species-specific poisons, and other chemical and cultural methods of preventing rodent damage to forests should be intensified.'" I wish he had stopped there instead of adding the following: "'Nevertheless, when damage problems are severe and rodent control is deemed necessary, the rodenticide described in our experiments is known to be economical and effective and affords some measure of safety to bobwhite quail.'"

I wish he had said something like this: Economy and effectiveness must no longer be measured in terms of money spent or earned by the company, by any company, but in relation to safeguarding the environment. While an individual who takes the wrong medicine, or even too much of the right medicine, is likely to harm only himself, the grower who uses the wrong poison or overuses the right one may very well affect an area far beyond his own fields. Certainly that was the case with Bowaters in Tennessee.

We are told we must have pest control to meet the needs of the booming human population for food, fiber, and protection from disease-bearing and nuisance insects. But this is precisely why we cannot continue to use ecologically crude insecticides in an inefficient, disruptive, and pollutive manner. Further disruption of the ecosystem, already badly bruised by harmful land-use practices and unrestrained waste, must end or we will no longer survive as a prosperous nation.

We need to kick the synthetic habit, the psychotic reliance on poison, and reject the old justifications given for it in the name of service to
humanity. Government agencies and state universities, above all, must
learn a new way of life and overcome their addiction to poison.

Several months ago I received an interesting report from the North-
east Forest Experiment Station exposing the old fallacy that reforestation
by direct seeding depends on the use of the chemical endrin. The report
showed that the best protection comes from careful planting, ensuring
that seeds are completely covered with a thin layer of soil, and also that
coated seeds sown on the surface suffer severe losses of 50 percent or
more, regardless of endrin concentrations as high as 4 percent. I considered this most encouraging and wrote the Forest Service hoping to
learn that such projects were going on elsewhere as well. But what I got
from the head of research was mostly a report on applying endrin better, particularly through impregnation into the seed, which supposedly
reduces loss of the chemical into the environment. I think he would
have said the Forest Service is desperate to get away from endrin, since
it tends to be magnified in natural ecosystems by being passed from one
link to another in food chains, and thus to poison the whole landscape.

In my own state, Virginia, a bulletin on "Weed Control in Pastures"
from the VPI Extension Service promotes the use of 2,4-D and similar
herbicides. But the bulletin is loaded with such cautions as "Small
amounts of these herbicides are capable of damaging or killing susceptible plants such as alfalfa, many clovers, cotton, tobacco, grapes, most
flowers and ornamentals." In the case of dicamba, it adds, "Do not graze
meat animals in treated fields 30 days before slaughter. Do not use seed
from treated grass for food or feed purposes. Do not apply on or near
desirable trees or plants. Do not graze dairy animals or treated animals
within 7 days." Clearly here is a poison requiring not just caution in use
but total elimination.

The poisons in this family are widely used in forest management.
The Bowaters people wrote to me in response to a request for information that as far as they know 2,4,5-T is not toxic to any form of animal life.
And Westvaco advised, "We have made it a practice to observe carefully
the results of these applications and have not detected any adverse
effects."

Someone should tell these companies the facts of the case. The dangers in 2,4,5-T are best known, largely through the tragedy of Vietnam.
Yet 2,4-D is the most widely used pesticide. About 79 million pounds

were produced in 1968, more than 20 percent of the total herbicide production of about 375 million pounds.

Dr. Samuel Epstein, of the Children's Cancer Research Foundation, declared before the Senate Subcommittee on Energy, Resources, and the Environment on April 15, 1970:

> Continued use of these herbicides in the environment constitutes a large-scale human experiment in teratogenicity. At its worst, the experiment may produce effects on human infants that become part of the overall pattern of birth defect pathology. Such effects probably may not even be recognized for what they are, at least for some time, since the significance of their incidence may be obscured in the overall and inadequate statistics on birth defects. Such an experiment is unwarranted by any conceivable criteria in the face of the unambiguous warning sounded by available scientific data.

Restrictions are continually being placed on the use of these poisons. They should be banned throughout the world until more is known about their dangers. And when in doubt, their use should be stopped voluntarily.

"What are Pesticides Doing to Human Beings?" is the title of a booklet published by the American Nutrition Society, advising which foods are safe and how to prepare them: "Peel all fruits and vegetables that lend themselves to such treatment. Scrape the outside layer from celery. Washing will not remove residues. Purchase eggs from local ranches not using DDT, BHC, or chlordane in hen houses." The Mississippi Game Commission reports that "long lived insecticide residues are causing portions of life to either disappear or evolve to an extent that may render them harmful to man and other consumers." Approximately five thousand reports of pesticide poisonings are received annually, but officials of the National Clearinghouse for Poison Control Centers believe the total number of such poisonings is eight to ten times greater. An official of the Michigan Department of Natural Resources urges leaders in his field throughout the nation "to fight the system of regulative permissiveness towards hard pesticides and the steamroller tactics of the industry which continues to push these economic poisons." This feeling goes right to the grass roots. "They are continually spraying the area," complains the president of a hunting club in Tennessee, "killing hard-wood, all types of insects, bees, butterflies, hummingbirds, etc. We believe these poisons are absolutely a detriment to the health of human beings and all types of wildlife."

The worst notwithstanding, I look hopefully to stirrings of con-science and responsibility. It is encouraging to learn of experiments with potted plantings, or tublings, instead of endrin-coated seedlings. These may be economically motivated, but they demonstrate the answer lies in better management techniques rather than poison. I read an interesting piece in the September 1970 issue of the *Journal of Range Management:*

> The volume of literature on the use of herbicides on rangelands is quite large and growing. In this body of literature it is rare indeed to find any consideration or discussion of the ecological impact of herbicides on the environment. It is relatively safe to say that effects on non-target flora and fauna have been of little or no concern. . . .
>
> Spraying and artificial treatments that reduce diver-sity can seldom be justified. Diversity and variety are cornerstones of survival since the more ways there are of consuming, or being consumed, the more favorable the chances of a living population to avoid high fluctu-ations in the birth and death of individual species. Agri-cultural practices that reduce the diversity of species in a community may be working in exactly the wrong direction. . . .
>
> Range management as a science will not survive unless it adopts a meaningful philosophy and under-stands that any science that relates to the land resource cannot stand alone, or isolate itself structurally by con-centrating within a narrow single use field.

Certainly the same applies to forest technology. Take it another way from James K. Lewis in *The Ecosystem Concept in Natural Resource Management:* "A high degree of human control over range ecosystems is usually either not possible or not economical. If a high degree of human control is economical the land is usually cultivated and ceases to be range. Consequently, range must be manipulated by extensive methods which are ecological in nature rather than by intensive methods that are agronomic in nature."

Thus, you see, I am not pressing for biological controls, which are much under discussion these days. The answer lies not in technical specialization, but in the objective of our society. The technology we develop must be governed by ethical principles. Then we can properly determine where to use what and how to apply it.

Based on thirty years of personal experience with the gypsy moth problem in Ulster County, New York, including his own family holding of seventy-five hundred acres, Daniel Smiley advises as follows: Knowledge about the moth and its control is greater than ever before. Valiant and honest efforts have been made by federal and state technical people to control the insects. Great sums of public money have been spent. And gypsy moths are present in a wider area than ever. There was no spraying of poison on his place, yet the gypsy moths died, he believes, from the combined effects of disease (probably wilt) and parasites (wasps and flies). The forest leafed out again—it was a case of true biological controls, without human help except that of letting nature take its course.

[The gypsy moth was deliberately imported from France in 1869 by an enterprising entomologist who wanted to interbreed it with silkworms. (Another strain also arrived from Asia.) But the gypsy moth spread its wings, took flight, began chewing leaves on trees and became a problem. In 1953 state and federal agencies began using DDT to control the gypsy moth, but that proved a bad idea since DDT got in the soil and showed up in cows' milk.

[The chemical crusade against the gypsy moth has been waged with publicity and promise. In the spring of 1966, the Michigan Department of Agriculture distributed a brochure titled *Oh Where Oh Where Did the Gypsy Moth Go?* which reported on the supposedly successful million-dollar eradication program. But in 1973 a new round of treatment pledged "reducing gypsy moth populations in Michigan to non-detectable levels." In 1992 the Forest Service and its state cooperators initiated the Gypsy Moth Slow the Spread (STS) Project. Its aims include avoiding "at least $22 million per year in damage and management costs," even while gypsy moth control is like a growth industry.

[And now the same forces that brewed the chemical feast have prepared a delectable new platter called Biotechnology, heralded as the way to feed the world—through factory farming, of course. It advances agronomy to higher levels of manipulation and control of nature. But thousands of varieties of genetically engineered organisms are likely to pose greater risks than benefits—risks in loss of diversity, contamination of conventional crops, development of antibiotic resistance in bacteria, and the emergence of new diseases in plants and animals.

[Some of forest technologists for their part are eager to team up with biotechnology and replace natural forests with genetically engineered tree factories, whole plantations cloned in monoculture. Those plantations would be increasingly susceptible to disease and to resistant "super-pests." They would be dependent on chemicals that kill beneficial insects and poison wildlife. Just as hatchery-bred fish affect and degrade wild native stock, the pollen of engineered trees is likely to degrade surrounding wild forest.]

Environmental forest management dictates that values of timber yield must be balanced with protection of soil, water, wildlife, and scenery, and with assurances that harvested areas will grow more trees for future timber needs. The mixed hardwood or hardwood-and-pine forest is a complex, diverse, and relatively stable association of plants, with a tendency to maintain its ecological norm. There is plenty of room for manipulation within the norm, along with growing timber on long rotations. In a closely managed forest, sapsuckers and other birds by their activities will indicate which trees should be culled. Drastic changes outside the norm and the widespread conversion of hardwoods to pine may be efficient in terms of technology and short-range industrial economics, but they are likely to be disastrous in the long run.

Farm-lot forest management favors the irruption of pests and disease. Infection is rapid and direct from tree to tree. If one species is destroyed there is nothing left. A monoculturally managed forest therefore creates the need for herbicides and pesticides, which further destroy diversity, including the soil nutrients, leading to unending applications of chemical poisons and fertilizers and steadily decreasing productivity.

In summary, the more complex and diverse a community, the greater its stability. That is basic ecology and good sense. To worry about deer or board feet, or even endangered large carnivores like wolves, is not enough. Some four hundred thousand kinds of plants, some two million kinds of animals offer an unending variety of form and function, of unsolved biological mysteries and rewarding beauty. Take away the diversity and man is lost. "The prime concern of mankind, from now until the end of human time," according to Dr. Hugh Iltis of the University of

Wisconsin, "will be the conservation of the diverse biotic environment, the only one to which man is adapted, the only one in which he can live."

Stated another way, any decision to inflict death upon another creature needlessly or wantonly is a decision to move toward death oneself. The Society of American Foresters ought to be committed to a better idea.

■　■　■　■

In 1970, when I gave this speech, Hugh Iltis was already well recognized as a scientist with strong social views. He became for me a scientific consultant, then ally and friend. He was, by his own confession, an "addicted botanist," or "half-plant." For many years he was professor of botany and director of the herbarium at the University of Wisconsin until retiring in 1993 and becoming professor emeritus. Often he would stress to me the transcendent importance of knowledge, science, and data, as though I greatly needed the lesson. Nevertheless, in the *Atlas of the Wisconsin Prairie and Savanna Flora,* which he co-edited (with Theodore S. Cochran) in 2000, he wrote a personal essay entitled "Humans and Mother Nature, the Unbreakable Bond," that concluded as follows:

> Prairies, as much as tropical rainforests, are part of our holy Mother Nature, and we neglect her protection from humanity at our very own peril. Think globally, but act locally, if not for the sake of our prairie flowers, at least for the sake of our own children. . . . If we succeed, we can have hope that children all over the world, will have a Mother Nature they can call their own, and that in Wisconsin they will be able to lie quietly in the grass on a sunny prairie hillside filled with flowers, watch bumblebees visit shooting-stars and pasqueflowers, hear dick-cissels and meadowlarks call in the sky, and be ever enchanted and empowered by that great symphony we call life.

Prejudice, Predators, and Politics

At the Great Plains Animal Damage Control Workshop,
Kansas State University,
Manhattan, Kansas,
December 11, 1973.

Everywhere in America during the 1960s I found blight freshly spread across the landscape, and that was before the ascendance, or transcendence, of the golden arches. Peter Blake, then editor of *Architectural Forum,* wrote a book, *God's Own Junkyard,* showing in all its glitch the modern outdoors décor of billboards, neon signs, automobile junkyards, and uncontrolled ribbon development, degrading the landscape and public taste.

I was particularly concerned at first with touristic aspects—the sleazy tourist traps, commercialized nature, phony Indian trading posts, and roadside zoos. The Black Hills of South Dakota provided a classic case. The twenty-five-mile road from Rapid City to Mount Rushmore was an obstacle course lined with "family attractions" like the Reptile Gardens, Gravity Hill, Gravity Spot, Dizzyland U.S.A., caves, gold mines, and Fairyland's Bewitched Village. The more I saw, however, the more my interest focused on animals caged at souvenir stands and gas stations, and on animal "farms," animal "gardens," and animal "parks," virtually all of them displaying scraggly and mistreated deer, snakes, and bears. They were the poorest places to introduce children to wildlife.

One of the worst single exhibits was the caged bear at the historic gateway to Pisgah National Forest, in western North Carolina. Tourists were invited to buy a Coke, hand it to the bear through the bars of its cage, and watch the bear drink it. Another like it was located at Cherokee, on the Indian reservation at the entrance to Great Smoky Mountains National Park. I wrote critically about them and pressed the North

Carolina Wildlife Commission for comment. It conceded the bears were kept in "deplorable, unsanitary and inhumane conditions" but didn't do anything about it.

Bona fide tourist enterprises may have been embarrassed but were not disturbed. Local newspapers didn't write about these things, and neither did visiting travel writers. I felt someone should and went around knocking on doors of wildlife conservation organizations in Washington. The National Audubon Society, Humane Society of the United States, and National Wildlife Federation all were sympathetic but preoccupied with what they considered higher priorities. I corresponded with the American Humane Association, which was too busy, too, but asked me to be sure to mention it in anything I wrote.

Somehow I wandered into the office of Defenders of Wildlife in the Dupont Circle Building, a great old building well located in downtown Washington, honeycombed with cubbyhole headquarters of do-gooder organizations. Mary Hazell Harris, a gracious, aging lady, the executive director, and the whole staff of Defenders, gave me a warm welcome as though glad to have company. She listened to what I had to say and invited me to write an article about the caged animals and roadside zoos for the Defenders' magazine, which she edited.

Defenders was not an outstanding magazine. It was not professionally edited. I didn't get paid for my contribution, but I was glad to do it. Part of my reward was to reach a sympathetic audience and another part was to learn what Defenders was about: It was an uppity little off-brand outfit started in 1947 as Defenders of Furbearers by scientists and esthetes (including Devereux Butcher of the National Parks Association, whom I have mentioned elsewhere) to represent the cause of coyotes, wolves, bears, and other predators. It was new to me, but it made sense.

In due course I wrote "Predators, Prejudice and Politics," one of my first articles for *Field & Stream* (December 1967). In preparing that article I asked a lot of questions and was dissatisfied with the inadequate answers I received. Predator control was conducted aggressively by the Fish and Wildlife Service, an otherwise weak-sister agency of the Department of the Interior. Secretary of the Interior Stewart L. Udall commissioned studies and pledged improvements, but mostly for public relations. Nothing really changed, but then Udall was an Arizona politician, and in Arizona predator control was a political reality.

Things did change under Nixon. Walter J. Hickel, Nixon's first secretary of the interior, related that while camping deep in Hells Canyon on the Idaho–Oregon border, he was met unexpectedly by a very aggressive sheep rancher. "Mr. Hickel, I lost seven hundred lambs to eagles last year. What do you think of that?"

"I think I'd get the hell out of the sheep business. . . . If he had said 'seven' I might have been willing to listen. But seven hundred—ridiculous!"

Nixon dumped Hickel for other reasons, but even so, in his 1972 environmental message declared, "The old notion that 'the only good predator is a dead one' is no longer acceptable as we understand that even the animals and birds which sometimes prey on domesticated animals have their own value in maintaining the balance of nature." Neither John F. Kennedy nor Lyndon Johnson had said anything like that. Then Nixon showed he meant it by issuing an executive order banning the use of poisons against predators on all public lands except in emergencies. He also asked for legislation to transfer predator control to the states, to provide cost-sharing funding to states not using poisons, and to fund research into environmentally sound predator control.

The sheep lobby and other beneficiaries of the control program blocked passage of the legislation by Congress and induced Ronald Reagan as president to rescind Nixon's executive order. The program ultimately was transferred to the Department of Agriculture, and all these years later, into the next century, the senseless killing goes on.

Thirty years ago officials of the Fish and Wildlife Service demeaned national parks and wilderness as "reservoirs of predation" that threaten game species, thus reflecting the prevailing point of view in wildlife management. And in the year 2000, when the same federal agency proposed to reclassify the status of the wolf in the northeastern United States under the Endangered Species Act, it downlisted the species from "endangered" to "threatened." That was plainly a political means of catering to state wildlife agencies, since threatened status would give them "management flexibility" to destroy wolves should they be perceived to be taking too many deer and moose, their natural prey species. Then in 2001 the Oregon state wildlife agency announced a plan to kill mountain lions to test whether the lions were reducing elk herds, while the Colorado wildlife agency was considering a ten-year study to kill coyotes to protect mule deer. That is how the state agencies and wildlife managers in charge view their responsibilities.

Wolves have come back. In 1969 I wrote a book for children, *The Varmints: Our Unwanted Wildlife,* in which I recorded there were then less than five hundred gray timber wolves in the entire United States outside of Alaska. Wolves were found in only one national park, Isle Royale, an island in the Michigan portion of Lake Superior, and in only one national forest, the Superior in Minnesota. They were not found in a single wilderness area of the West.

Today, early in the twenty-first century, wolves are widespread, if not abundant. They have a rough go in the Southwest, where the political power of the livestock industry has inhibited reintroduction of the Mexican gray wolf. Elsewhere, however, campers and backpackers can listen to the long-drawn musical note, the vast wail with echo and reecho that we call the "wolf howl," and learn lessons from an animal of endurance, courage, and loyalty to its own kind. The return of the wolf is one of the major environmental and ethical achievements of American society during my lifetime. It results from the manifestation of concern and demand by many, many people, and by organized groups like Defenders of Wildlife. Some wildlife management professionals have been part of this movement, but certainly not their agencies nor their establishment.

I'm glad I wandered into the office of Defenders of Wildlife and advanced from dealing with roadside zoos to larger issues. In due course I was elected to the board of directors, although I'm not much into going to meetings. Later an eccentric tycoon willed part of his fortune to Defenders. It became a bit of a different organization, with officials, expense accounts, and secretaries. Still, after I was dismissed from *Field & Stream,* the editor of *Defenders* called and said, "I'm authorized to offer you space for a regular column—as long as you don't get us sued for libel." I accepted and that went on for eighteen years.

■　■　■　■

Indiscriminate trapping, shooting, and poisoning have reduced some of the rarest, most beautiful, and superbly adapted species of our wildlife heritage to the brink of extinction, although they constitute a resource that could be enjoyed by all and harvested by sportsmen under sound management principles. The war on predators has been waged with little scientific knowledge of their beneficial role in the biotic community, and without moral or ethical consideration for man's responsibility in preserving natural life as an integral part of the environment.

The Division of Wildlife Services, an agency of the Interior Department, has had one prime goal at the root of its existence: to kill wildlife.

It has for years gotten away with murder—the murder of wolves, mountain lions, coyotes, bobcats, foxes, badgers—as well as anything else that might be handy.

For years sportsmen were led to believe that elimination of predators would result in an increase in game. Certainly a given range will support only a certain number of animals, whether game or domestic stock; but, as we have learned, predators take only small numbers from the animals they prey upon and are probably essential to maintenance of a healthy, virile population. These lessons were made abundantly plain through scientific observation of moose and wolves at Isle Royale National Park in Michigan. Wolves were noted to claim the old, diseased, heavily parasitized, and weak young among the moose. As a result of natural selection, the closely cropped herd is healthy and among the most productive on the continent, bearing a high proportion of twin calves.

Isle Royale may be isolated, but principles learned there apply to predator-prey relationships the world over. Hunting plays an important role in helping to remove excess population, but, unlike natural predation, hunting does not select the weak unhealthy specimens. When predatory population is excluded from a natural community the weaker members remain, weakening future generations of the species. But natural predation keeps the old and weak individuals to a minimum, benefiting the hunter as well as the herd. Predation maintains qualities of wildness in game species that we admire. Fleetness, grace, alertness, and deception give the feeling of challenge and pride not found in killing a chicken or cow. The animals that have these qualities are the product of centuries in their struggle of survival from predation.

One of the finest achievements of the Nixon administration in the resource area has been its positive program to bring poisoning to a halt and to restrain the Division of Wildlife Services in its operations on public lands. Until now there has not only been an abrogation of responsibility in diverting public funds to serve a special interest, but that interest, the sheep industry, has caused more damage to the land through the years than almost any other force. Overgrazing, first at lower elevations, then on alpine meadows, has ruined millions of acres, removing plant cover, disrupting animal communities, and devastating watersheds.

Overgrazing sets vegetation back to early stages. On such misused land, ground squirrels, rabbits, woodrats, hares, and pocket gophers are abundant, very probably contributing to rising numbers of coyotes. Beyond a doubt, the first consideration in "animal damage control" is good land management. This doesn't mean the elimination of livestock on public lands, but it does mean serious restraints to ensure that stock is brought within the carrying capacity of the land. Unfortunately, sheepmen are incredibly ill informed on these questions. They grasp at straws to hold back the great movement toward environmental ethics and morality. Some sheepmen haven't the foggiest notion of wildlife ecology, or animal behavior, though they have lived in the company of animals all their lives.

It is essentially significant that this animal control workshop should be held here in Kansas, where control is conceived as part of management to help the landowner solve his own problems through sound conservation practices. Through the extension system, farmers and ranchers are shown how to concentrate on catching the individual coyotes and other predators guilty of killing livestock, while those without damage do not waste their time and money chasing down imaginary or harmless predators. Publications and films explain predation as a necessary, beneficial part of life. The federal government has no documents like these.

The nation has a long way to go toward sound management of its predator resource. Prohibition against the use of poisons would be one positive step in the right direction. Acceptance of the extension system by more states would be another. So would placing hunting of predators on a sustained-yield basis. All these are in keeping with a change in values worthy of an environmental age.

Protecting the Public Options

At the Fontana Conservation Roundup,
Fontana, North Carolina,
June 1975.

The Great Smoky Mountains and southern Appalachians around them in 1975 were still deservedly renowned for their scenic beauty and for forests and streams supporting one of the most diverse collections of plant and animal life in the world. Little wonder that here was the most popular, most heavily visited, of all national parks.

When I delivered the keynote address at that year's Fontana Roundup—a kind of inclusive, good-feelings regional conservation assembly—I spoke critically, but nowhere near critically enough. I failed to warn that magic mountain vistas were being diminished and degraded by smoke and haze, or "smaze," the result of toxic emissions airborne from factories and power plants well outside the park and from automobiles driven by tourists passing through it.

Since then, the high forests have become tragic catch basins of acid rain, seriously weakening the stands of balsam, Fraser fir, and red spruce that visitors come to enjoy. The Great Smoky Mountains now has the distinction of suffering the most polluted atmosphere of any national park. And those atmospheric pollutants eventually get into streams and lakes, destroying or impairing aquatic life, placing the entire ecology of the park area under severe stress.

The Smokies and surroundings in the early twenty-first century constitute a health hazard zone with high ozone levels and ozone alerts during the warm-to-hot summer travel season. The mountains at times are shrouded with smog—with loss of up to 70 percent of visibility. Mount Mitchell State Park, embracing the highest mountain in eastern United States, is critically affected, reduced to harboring a dead and dying forest.

Considering that tourism and recreation are major factors in the economy of the region, these natural resources logically ought to be safeguarded like treasures and restored where damaged. Environmental organizations such as the Western North Carolina Alliance persistently call attention to the devastation, but few in commerce or in politics respond with anything more than lip service, and certainly not with sincere remorse or stewardship. But, of course, tourists still come, unaware of nature's unhappy condition, enabling business to go on as usual, and allies of business in politics to damn restrictive government regulations and zoning.

I revisited Asheville in the fall of 1998, flying there from Morelia, Mexico, an ancient place grown into a city of one-million-plus, laboring under a thick, dark cloud of polluted air. In Asheville, it was much the same; I felt I had hardly left Mexico. But I remembered Asheville half a century earlier as a charming, healthful mountain resort. Wasn't that what led people to move here in retirement from New England, the Midwest and Florida? But who was minding the store to protect the qualities they came for?

"The preservation of scenic vistas signifies the commitment we've made to preserve the natural beauty of North Carolina," Governor Jim Hunt declared In 1999. "Once these areas are allowed to be developed, their priceless value is lost forever." The governor was absolutely right. He was referring to acquisition by the state of the prized forty-four-hundred-acre Lake Logan tract in Canton, between Asheville and the Smokies, from the Champion Paper Company. But the state, federal agencies, and local communities have not done nearly enough to protect the rural mountain landscape, or the urban landscape, for that matter, and even less to protect the air quality.

This is all part of a trend. In the late 1990s the Cherokee welcomed the gambling casino that now is the biggest enterprise on the reservation, far overshadowing the classic outdoor drama "Unto These Hills" and other traditional heritage presentations. And in the last days before Congress adjourned in the year 2000, Senator Jesse Helms and the local congressman, Charles Taylor, two of the most reactionary, anti-environment members of Congress, attached a rider to the highway appropriations bill for construction of the controversial twenty-one-mile road from Bryson City to Fontana Dam that would destroy the largest de facto wilderness within the national park.

I suppose in a· democracy we the people get no less than we demand and deserve. On the positive side, the New River, which I cited in my 1975 Fontana speech as a struggle in progress, did indeed become part of the Wild and Scenic River System; the Tennessee Valley Authority did not build its proposed dams on the Upper French Broad; the Tellico to Robbinsville road across the national forests was not built; and at the outskirts of Asheville the North Carolina Arboretum now covers 426 acres of gardens, greenhouse, and educational displays as "a natural cradle of plant cultivation." For the hopeful, there always is hope.

■ ■ ■ ■

It is always a pleasure to come to the Smokies and the surrounding Appalachians in order to review the scene of *Strangers in High Places* and to be reminded of how glorious these southern mountains really are.

I can't help but pay tribute to two men who helped shape my appreciation of this region and who inspired me to interpret and to defend its natural values to the fullest of my ability. One was Garth Cate, of Tryon, the founder of the Fontana Roundup. When Garth died, Mrs. Cate wrote me that he had felt as a father to me. I was obliged to reply that I never felt as a son to him for the simple reason that Garth's outlook was ever young. He may have been much older in years, but certainly not in spirit. Garth was over ninety when he died, but thinking and pointing to the future, rather than ruminating over the past. [I met Garth Cate when he was in newspaper promotion work in New York. Then he became a consultant in the travel industry, emphasizing to resorts the values of landscaping and beautification.] Harvey Broome, of Knoxville, the other source of inspiration to whom I refer, was much the same way. He was a gentle, selfless, and thoroughly determined man. Harvey was president of the Wilderness Society at the time of his death, with far-reaching interests and activities, but his heart was in the Smoky Mountains. I daresay that the security of the wilderness within the national park at our doorstep is due to the diligence and devotion of Harvey Broome more than to any other man.

Before heading down from Washington, I thought I should go to the Capitol and obtain clearance from Representative Roy Taylor to come into his district. [Roy Taylor, a Democrat, had a measure of national prominence as chairman of the parks subcommittee of the House Committee on Interior Affairs.] Mr. Taylor examined and did indeed find my papers in order and extended freedom of passage in western North Carolina. We have had our differences, but I have always respected the congressman as a competent legislator of integrity and principle.

I recall once, several years ago, Luther Shaw, his administrative assistant, said to me, "Now listen, Mike, you and the environmentalists have us blocked at almost every turn: You have stopped construction of the transmountain road in the Smokies, the Tellico Plains–Robbinsville Road through the national forests, the TVA dams on the Upper French Broad River, and Beaucatcher Mountain tunnel in Asheville. Why don't you be reasonable and let us have something?" Actually, I haven't been involved in the Beaucatcher project at all and feel unqualified to comment. On the other three, however, I can speak with some experience of involvement. In all cases it seemed hopeless to resist because they had been planned by powerful forces and their completion seemed inevitable. Nonetheless, when there's something worth saving one must never give up, certainly never give up hope. Because enough people cared and refused to quit, these programs for land desecration have either been scrapped completely or remain unfulfilled to this day.

There is always hope as long as there is a cadre of people, or even a single individual, with commitment and a willingness to fight. There is hope at this very moment that we may still save the beautiful New River in northwestern North Carolina and southwestern Virginia. It hasn't been easy to overcome the power of the utility companies and of the Federal Power Commission. Congress in the last session might have given the New River protection of the Wild and Scenic Rivers Act, but surrendered to the iniquitous nationwide lobbying of the utility companies and their allies in the high command of the construction unions. At that point all seemed lost. But the leaders in the government of North Carolina, responding to the wishes of the people of this state, have consistently pursued their legal appeals to save the New River. Moreover, the North Carolina legislature has voted to designate 26.5 miles of the river as part of its Natural and Scenic Rivers System. This in itself does not save the New River, but the secretary of the interior can designate state-administered components of the National Wild and Scenic Rivers System. Governor Holshouser has made application that this be done. North Carolina environmentalists have done their share. Now it is up to the rest of us to encourage the new secretary of the interior, Stan Hathaway, to respond positively—to recognize an opportunity to get off on the right foot, following the abundant criticism and mistrust surrounding his nomination.

North Carolina is moving on several fronts to achieve protection and sound management of its land and water resources in ways that other states could afford to study and follow. The legislature has passed an act barring billboards within one thousand feet of the Blue Ridge Parkway, an action urgently needed to preserve the most scenic corridor in America from commercial blight. Now the legislature is considering the Mountain Area Management Act (following passage of the Coastal Management Act of last year). Maybe action hasn't come as fast as it should, but anyone can see that the upland meadows, fields, forests, and farms are gradually being supplanted by subdivisions, tourist attractions of varying quality, and assorted other commercial enterprises. People of the mountain areas, not only in North Carolina but everywhere, will have to come to grips with some form of regional land planning. Local and state bodies must give serious attention to cooperative planning relating to land use and economic development. Strong citizen support must be solicited and encouraged to prevent further exploitation and unplanned growth from seriously damaging the highlands environment and from destroying its natural and recreational integrity.

We citizens will need all the hope and determination we can muster. We are now being robbed of our land-use options. At a time of critical resource shortages, big industries are pressing to maximize immediate profits without regard for long-term consequences to the land. President Ford and his administration are giving them all possible encouragement and assistance. No other president has made as many wrong moves and continues to do so as Gerald Ford. He has given absolutely no recognition to the truism that proper management of limited natural resources and protection and enhancement of the quality of life are investments of economic value for the future of America and the world.

When he became president, Mr. Ford spoke of establishing a mood of "communication, conciliation, compromise and cooperation." His appointment of Stan Hathaway, a hardline anti-environmentalist if ever there was one, showed that Mr. Ford has no desire for meaningful communication or cooperation with citizens who care. [Hathaway was appointed secretary of the interior in June 1975 but didn't stay very long, choosing to resign in October of the same year.]

The president has given no leadership to meeting the desperate need for energy conservation, no answer to unrestrained growth mania. At our present rate, the shrinking resources of our land are doomed. All sections of the country are under attack—Appalachia, the western plains, the Four Corners in the Southwest, the outer continental shelves, and the Arctic slope of Alaska. The public's land-use options are being rapidly preempted for the benefit of private profit. Clearly, we must begin at once to alter the life-style that makes us enemies of ourselves. We must reassess the value system by which we confuse superconsumption for the quality of life.

Land-use regulations are critically needed to protect what options remain. The nation can move toward sensible regulations through passage of H.R. 3510, now pending before the House of Representatives in Washington. This legislation does not create federal zoning, threaten private property, remove decisions, or provide sanctions. It does furnish grants to states that help them in their own planning process, based on simple requirements for preparing and administering plans. It does provide protection of "areas of critical environmental concern," but even this is left to the states' discretion.

H.R. 3510 has been the subject of heavy lobbying by the agents of the timber trust and the U.S. Chamber of Commerce, who have distorted the issues [and defeated the bill]. The fact is that federal policies already have a major impact on land use; legislation such as H.R. 3510 would actually restore power to state and local governments by helping them to solve major land-use problems themselves.

As an environmentalist, I share the apprehension manifest at grassroots America over the spreading influence of federal agencies. I can't think of one single agency that is tuned into the desires and needs of the people and truly responsive to them. Assorted land-management agencies may be represented at this Conservation Roundup, but they are here to tell us all the good things they are doing, not to find out what we really want them to do. I speak specifically of the Tennessee Valley Authority, Corps of Engineers, Forest Service, National Park Service, and Fish and Wildlife Service. Fortunately, we have means of restraining these agencies from abusive land-use practices—other than the pending H.R. 3510—such as legal action in the courts, the pressure of public exposure through the media, the education of the people as to the damage to the public estate being conducted with our own funds, the passage of new laws (such as the Wilderness Act, National Environmental Policy Act, and the Endangered Species Act), and congressional transfer of land from one jurisdiction to another (as in the case of the new North Cascades National Park in Washington state, formed from former national forest land).

The Forest Service, in particular, is doing intensive damage to the resource while conducting a charade of public involvement. In past years there were the "show-me trips" and "listening sessions." The "listening" was accompanied by a stream of color brochures and other promotions extolling clear-cutting, a thoroughly "efficient and economic system" of placing timber production above all other values. Cutting soared, tripling in volume over a period of twenty years; the resource, in turn, suffered, and wildlife habitat deteriorated.

"Public comments are invited but the consultant atmosphere appears to be lacking," complained Senator Jennings Randolph of West Virginia in early 1972, after an investigation at the behest of his constituents. "The prevailing feeling expressed by those after attending the hearings is that decisions have already been made and their expressed concerns have only been accepted as an empty polite gesture." [Senator Randolph was referring to the long, bitter controversy over management of the Monongahela National Forest in his state.]

In the same year, Charles Prigmore, president of the Alabama Conservancy, charged much the same: "The Forest Service indeed holds hearings at infrequent intervals, particularly when public pressure becomes irresistible. But this is an attitude of patient tolerance of public concern and thinking, rather than any real encouragement of joint decision making. Forest Service personnel consider themselves to be the experts, and the public to be ignorant at best and obstructionist at worst."

Public involvement, understanding, and support are essential to any land-management agency genuinely committed to leadership and service in behalf of the people. On this foundation the Forest Service was born and blossomed many years ago. The alternative to involvement and alertness of the public is surrender to the special interests, the exploiters who never sleep and are unrelenting in their political pressures. This was perfectly plain during the fight over the National Timber Supply Act of 1969. "Believe me, we have waste of great magnitude in our national forests, parks and wilderness areas," Jim Bronson declared on behalf of the National Forests Products Association in the midst of that fight. If the people had not risen to defeat that act, Mr. Bronson and his crowd would now be devouring the whole domain of public lands.

In 1970, the regional forester in the South, then Ted Schlapfer, sent me a letter enunciating his principles of public involvement. He wrote, "We are hired as professionals and paid by the taxpayers to make and act on decisions based on our education, training and experiences. While the public must be involved if we make changes in management practices, we still have the managerial responsibility of making the final decision."

Those days are done. No agency can be allowed any longer to make the final decision. Personnel in charge are too narrow in their training and experience to be considered sufficiently competent. And in innumerable cases they have shown a lack of forthrightness worthy of trust. This is not my own idea alone, I assure you, but an increasingly widespread feeling among concerned citizens everywhere. According to a recent issue of the *Georgia Conservancy Newsletter,* the Conservancy's executive committee has voted to challenge Forest Service planning and management of timber operations. The Conservancy won't even trust the foresters as timbermen.

The Forest Service is so driven by its timber-first policy that it lacks the breadth to appreciate or properly manage the rest of the resources. Let me cite my experience in trying to get straight answers from Forest Service leaders over two issues in this very area: sacrifice of the Bemis tract in the Snowbirds and Little Buffalo drainage of Graham County here in western North Carolina, and construction of a portion of the Robbinsville–Tellico Road across the Falls Branch Scenic Area in the Cherokee National Forest, in Tennessee.

The former Bemis tract consisted of fifteen thousand acres of high mountain land privately held within the authorized boundaries of the Nantahala National Forest. When the lands became available for sale in 1971 the Forest Service made no serious effort to acquire them. This was astonishing, particularly in light of a statement made to me by Chief Edward P. Cliff stressing the need of large scale acquisitions in the Appalachians because of encroaching land-use conflicts. Moreover, the area embraced choice trout streams and a connecting link on the migration route of the last and largest bear country in North Carolina outside the Smokies.

When I inquired of the Forest Service for the facts about its delinquence and disinterest, I received the most curious replies. For instance, the forest supervisor in North Carolina, Del W. Thorsen, wrote me a letter which included the following: "As to foreseeable watershed impacts from impending developments, we can only surmise. We do know that the soils in Little Buffalo and Snowbird are relatively stable and are not highly erodible." Now how did the foresters know this about the soils, since no detailed soils study had been completed or was available to them? But such language flows easily in the typically bureaucratic brushoff given to concerned citizens. To continue: "Since most of the tracts were sold to individuals who presumably intend to build vacation cottages in a forest setting, we believe they will want to protect the environment and not damage the watershed. Construction of access roads possibly will expose more soil to the elements than any other anticipated use. However, this damage need not be long lasting if the roads are properly designed and maintained." No need for a Mountain Area Management Act there!

Then there was another letter from Deputy Chief E. W. Schultz, which I also found unsatisfactory. He wrote, "With respect to the auction, the Forest Service is without authority to participate in a sale procedure of this sort where payment is required prior to examination and approval of title by the United States. The matter was discussed at some length with the Nature Conservancy prior to the sale. It is our understanding that their decision not to participate in bidding for the property was reached after they had evaluated the situation."

This statement was an untruth, plainly and simply. I can say this on personal knowledge. On the day before the auction I was called in Washington by a Tennessee congressman, Lamar Baker. He was responding to pleas from sportsmen constituents hoping to prevent the Bemis tract from being cut into a thousand pieces. I urged him to call Patrick F. Noonan, president of the Nature Conservancy, which he did. Mr. Noonan, in turn, telephoned the Forest Service—it was the first contact on this issue—and was willing to make an effort to save the tract, even at this late hour; he was furious at the Forest Service disinterest, as he advised me later in the day. I recall in particular his grave concern over the potential precedent of the sale for owners of other large blocks of land in the Appalachian national forest boundaries. The Forest Service is not concerned, however, over the loss of bear habitat or trout streams, or rare flora, or the inevitable watershed damage, or pollution caused by roading and building in the high mountains.

Is it any wonder that citizens fought so hard for the Eastern Wilderness Act in order to take the basic decision making in land use away from the foresters? This process is still underway. Citizens are looking forward to hearings on S. 520, which would establish three additional wilderness areas and twenty-three study areas in the eastern national forests, and hopefully to passage in 1975.

[The Eastern Wilderness Act of 1975 in fact added sixteen national forest wilderness areas in the East, where only four had existed, and designated seventeen wilderness study areas to be protected as wilderness pending congressional review and action. The newly designated units included the Sipsey Wilderness, Bankhead National Forest, Alabama; Cohutta Wilderness, Chattahoochee National Forest, Georgia; Cherokee National Forest, Tennessee; and Otter Creek Wilderness, Monongahela National Forest, West Virginia.]

I'm not especially satisfied with the process of decision making or the policies governing land use within the Great Smoky Mountains National Park, either. The prevailing official policies are designed to cater to crowds of unlimited numbers rather than to preserve the natural resource for fitting use and enjoyment based on the carrying capacity of the land.

To quote from an article by Sam Venable in the *Knoxville News-Sentinel,* "Down below the Chimney Tops, work crews are busy with bulldozers, dynamite, chain saws and boom trucks. They're cutting trees, shooting rock, moving earth and hauling away river stones in order to carve a new trail up the mountain. Civilization is getting a stronger grip on the Chimneys." Apparently it never occurred to the park management to make its plans known in advance and to seek public opinion or comment. In fact, the way this came out was that someone on the park staff sent me a confidential message, I telephoned my friend Venable, and he proceeded to investigate on the ground.

The assistant superintendent [and later superintendent], Merrill Dave Beal, gave him this dubious explanation: "So many people are using the present trail, it's become necessary to improve it. In many places, the old trail is slippery because of seepage. In others, tree roots have become exposed due to heavy pedestrian traffic and erosion is taking place." If a preservationist on the public payroll would examine the same problem he might say, "It's time to reduce the impact of heavy use, or even to close the trail." After all, if the tree roots are exposed what's the point in blasting them out with dynamite or in trying to solve the erosion with a bulldozer?

But Mr. Beal, alas, explained it this way: "There are many people who simply can't get out and scramble over rocks to enjoy the park, but they still have a right to enjoy the Smokies. After all, it's their park, too." What an ingeniously dangerous statement! If the value of parks is to be measured on the basis of use by the greatest number of people, with the value of the implicit wild, natural resource, and the use of it by bears, birds, insects and plants, to be secondary, then all of our parks are doomed, one after the other.

It takes agencies repeated lessons before they learn. In 1974 the Park Service decided to proceed with the access road into Cataloochee Valley. In order to shortcut the requirements of the National Environmental

Policy Act, officials prepared an "Environmental Assessment" instead, based on their intent to proceed with construction within thirty days. They tried to keep it quiet, never notifying anyone on the Tennessee side, let alone any national publication, national organization, or any concerned citizens. Someone in Haywood County appealed to me out of desperation, and I did what I could to spread the alarm. It frightens me that this is how our public officials are doing business on our public land.

The Environmental Assessment arrayed valid reasons why the access road should not be built: a potential increase in use from 250 to 8,000 persons per day, a volume which the fragile valley cannot stand; potential for strong temperature inversions; inevitable soil erosion and stream siltation; inevitable destruction of wildlife habitat. Nonetheless, the agency decided to proceed. When I raised the question with the park superintendent of the potential for a temperature inversion to trap and concentrate auto emissions, as mentioned in the assessment, his response was, "That's just a potential. It doesn't say it's going to happen." If this is the kind of principle guiding our national parks, we are in deep trouble.

Nevertheless, the citizens were forced to take legal action in order to protect Cataloochee. Or perhaps I should say to protect the options in land-use decisions. How, indeed, could the citizens give any credence to the regional planning about to begin in the Smokies if the important options are foreclosed to them by road construction? Lately the Park Service has conducted a planning session on the Smoky Mountains region in conjunction with other federal, state, and local agencies. I was interested to read a report by John Parris in the *Asheville Citizen* where the multiagency task force of planners recommended that a comprehensive environmental impact statement be prepared by the Park Service on the proposed Cataloochee access road. That's not only a vindication of the citizen position, but a clear sign of the need for more heavy internal training by the National Park Service.

The guidelines of the task force also recommended that consideration be given to charging an entrance fee into the Great Smokies as an attempt to reduce congestion. Personally, I'm opposed to it. In the first place, it's illegal under the terms of the grant of land from the state of North Carolina to the federal government. In the second place, it tends to benefit those who can afford to pay while penalizing or discriminating against those less fortunate who cannot.

Most important of all, however, popularity and congestion are clear signs that the Great Smoky Mountains National Park needs to be enlarged.

The national park should be doubled in size, to at least one million acres, in order to meet the needs of the people in Appalachia and of the nation. One new section should include Mt. Mitchell, which deserves national park status in its own right as the highest peak east of the Rocky Mountains. The enlarged park should include the Nantahalas, which compare in magnificence to the Smokies themselves. Another section should include the unflooded Little Tennessee River, which TVA will have to surrender in order to comply with the Endangered Species Act, and the bordering historic lands that embrace Echota, the capital of the Cherokee nation, and the birthplace of Sequoyah. It should include the Tellico uplands as a national recreation area open to hunting, in the same relationship to the Smokies as the Ross Lake and Lake Chelan National Recreation Areas hold to the North Cascades National Park in the state of Washington.

Most of the lands needed for expansion of the park already are in public ownership, principally in the Pisgah, Nantahala and Cherokee National Forests. The summit of Mt. Mitchell is in a state park. The additional high-elevation lands should be acquired by purchase in order to ensure protection of the watersheds, of wildlife and fisheries habitat, and of the rare Appalachian flora. All these are implicit park purposes. [Representative Roy Taylor actually was sympathetic, declaring, "I've looked at a lot of areas and none was more majestic and beautiful than the Mt. Mitchell area." Before retiring from Congress in 1976, he obtained federal funding for a study of a potential new national park. However, the National Park Service showed no interest, and community hearings were dominated by expressions of hostility.]

Soils in the highlands aren't sufficiently deep or fertile over large enough areas to justify intensive timber management. But the qualities that make the area a liability for production make it a natural for human enjoyment: a variety of species virtually unequaled anywhere in the country, high ridges affording breathtaking views, high elevation affording cool climates and escape from congested cities and sweltering lowlands, streams for fishing and enough protective cover and food for wildlife. The land developers recognize these values and their marketability to those few who can afford it. Now, while the real estate market is down, marks the ideal time to acquire the land for the benefit of all people, for this generation and generations hence.

The multiagency task force has proposed establishment of a thirteen-county Great Smokies Recreation Region. This proposal is good, but not as bold or positive as the enlargement of our great national park. The purposes of the Forest Service and TVA are incompatible with the protection of unique natural systems in the region. The proposal in South Carolina to establish a Southern Appalachian Slope National Recreation Area, in order to protect a significant area of the Blue Ridge Front, which Representative James Mann is sponsoring in Congress, demonstrates the widespread citizen interest in natural area preservation. And there is little time to lose.

TVA represents an extremely serious challenge to sound land use. This agency, which began with great promise, has come to play a destructive role in the region with a negative impact on the land, its people, and its healthy development. TVA is presently unaccountable to the public, the public service commissions, or the states it is designed to serve. Public opposition is growing to TVA's huge electric rate increase, its promotion of strip mining and tall stacks, its proliferation of radioactive nuclear facilities, dams designed to wipe out beautiful natural streams like the Duck and Little T, and to self-authorized projects like the pumped storage development at Raccoon Mountain. TVA is the personification of bureaucratic arrogance. It is an agency that, like CIA, has grown bigger than the law.

Fortunately, the tide is turning. The Senate Public Works Committee recently held oversight hearings in Washington on TVA activities. It was the first time in TVA's forty-one-year history that such hearings have been held—and they can hardly be the last. Further, before leaving Washington, I learned that either this week or early next week the Department of the Interior will list the snail darter, a small fish found only in the Little Tennessee River, on the endangered species list. The department will declare the completion of Tellico Dam would "totally destroy the habitat of the darter." Since TVA has pledged compliance with the Endangered Species Act of 1974, it must voluntarily stop the dam project and stop it now, before any further damage is done.

[In 1975 the Interior Department declared that protection must be accorded to the snail darter on the Little Tennessee River under the Endangered Species Act. The validity of the act to save a seemingly obscure species was sustained in the Supreme Court and against attacks in Congress. In 1979, acting under the influence of Senator Howard Baker of Tennessee, Congress exempted Tellico from terms of the act; subsequently, the dam was completed and the Little Tennessee was flooded.]

Determinations of land use belong to everyone but above all to the little people. To quote a guest editorial appearing in the *Asheville Citizen-Times* on April 9, 1972, written by Merrimon R. Doster, executive secretary of the Franklin Area Chamber of Commerce,

> We cannot let ourselves be lured into swapping our quality of life for any quantity that could possibly result in any deterioration of our environment. If a new industry will increase incomes, raise living standards, strengthen our institutions, enhance the quality of life in general, then we want it and need it, but if it pollutes a stream—we can't afford it. If it dirties our air—we don't need it. If it defaces a mountain—we cannot tolerate it. If it reduces the dignity of our citizenry—we must fight it to its end.
>
> All our resources are natural—majestic mountains —pure water—clean air—grand scenery, and, above all, decent, intelligent people. Our people are the grandest resource we have, for it is within their power and choice to either conserve or destroy all the rest. . . .
>
> If we are to plan for the development of a tourist industry, we must first build the tools with which to work. The last tool to be used must be the first one developed. That is a method to stop when growth has reached that point of maximum efficiency, where resources are used but not abused. The cold hard fact is ever before us that when our resources are expended they cannot be replaced.
>
> That all-important tool is called "Land Use Planning."
>
> Land Use Planning is a difficult and ofttimes painful program to formulate and implement. But never have the penalties been as devastatingly tragic as those experienced in areas of unplanned growth. We must not make the mistakes that some of our neighboring states have made. We cannot afford to.

I view the future with a sense of great hope and patriotism. The Society of Colonial Dames has just given my new book, *Battle for the Wilderness,* an award of merit. I really didn't know they cared, but it makes me feel particularly patriotic. On the eve of the nation's bicentennial, we should commit ourselves at this Roundup to protect and perpetuate our most precious heritage, the American earth. Environmentalism, after all, is not concerned with solving problems of the present, but rather problems of the future. I can't think of a better legacy to hand down to the future than open options of land use.

Someone said to me here this morning that it's futile to fight the battle for the Little T because TVA has too much power and has foreclosed all opposition. Our forefathers of 1776 faced even greater odds. They were the masters of grass-roots activism. And now, all across America, the fight for a better environment has begun. Let us make the most of it.

It's Tough Enough to Make a Living, but Tougher to Say Something That Counts

On receiving the 1981 Award for the Best Magazine
Article of the Year from the American Society of
Journalists and Authors,
New York City, May 9, 1981.

I worked as a freelance writer for more than twenty years and made a decent living at it. Once, early in that period, I was offered a regular job on a magazine. My employer would have provided office space, telephone, stationery and business cards, postage, coffee breaks, expense account, a regular paycheck, paid vacation, sick leave, medical benefits, and retirement. I gave the offer considerable thought, prayed over it, and then declined with thanks. Freelancing certainly is a tenuous way to go, but I learned to cherish the freedom, challenge, and opportunity connected with it.

Freelance writers work alone, but we need each other too. They need to know that others are out there and to share the struggles and successes with them. As soon as I qualified, I joined the Society of Magazine Writers, later renamed the American Society of Journalists and Authors (ASJA), and, when I wrote my first book, the Authors Guild. The ASJA defines its role as offering "extensive benefits and services focusing on professional development . . . and, above all, the opportunity for members to explore professional issues and concerns with their peers." To my mind, that "above all" is above all.

Thus, in 1974, when I thought I was unfairly dismissed as conservation editor of *Field & Stream,* I wanted to make an issue of it and turned to the writers groups for sympathy and support. I had joined the staff seven years earlier as a contributing editor, which allowed me to pursue other interests as well. It was a bit of a strange connection, considering that I was not a hunter, nor much of a fisherman. My interest was in conserving, not in killing. Yet I wrote a column every month and two or three major features a year discussing environmental and ethical issues as I saw them for a large and significant audience that I found eager to receive the message and respond to it.

Unfortunately, in 1973 Clare Conley, the editor who had hired me, was let go and Jack Samson, his successor, was definitely not into environmentalism. He found the perfect rationale for getting rid of me in a citation from Cleveland Amory, an anti-hunting activist and a man the gun establishment loved to hate. Amory had lately published a book titled *Man Kind?* in which he quoted a page in my book *Battle for the Wilderness*, criticizing "slob hunters" and "jet-set gunners whose greatest goal is to mount on their walls one of everything that walked Noah's plank."

Perhaps I should have gone quietly, but many individuals and organizations protested my dismissal, and I received letters of support, faith, and confidence that gave me new strength and commitment. The ASJA Professional Rights Committee (chaired by Patrick F. McGrady Jr.) investigated my dismissal and issued a white paper highly critical of *Field & Stream*. No, it did not get me reinstated, but it showed that writers care about more than the paycheck, and that they care when one of their own is stomped on. Luckily, I did not die, I was only dismissed.

■   ■   ■   ■

I appreciate deeply this singular recognition and honor accorded to me by my peers and friends. I feel a strong sense of responsibility that my work in the future continue to justify the selection by the Awards Committee.

As for pride, I invite my colleagues to share the emotion of this moment with me. On first being notified by Grace Naismith of *Readers Digest*, the gracious committee chairperson, that I had been chosen as winner of the Best Article Award, I was carried away and offered to donate my five-hundred-dollar stipend to the society for use by the Professional Rights Committee. Then I had sober second thoughts, since I can use five hundred dollars as well as anyone. But the other night I was reviewing the records of my dismissal as conservation editor of *Field & Stream* in late 1974 and the protest that followed. I realized how much I owe to ASJA and its members who stood behind me. I might not be here today if it were not for Patrick McGrady, Marvin Grosswirth, Dodi Schultz, Mort Weisinger, and many, many others in this room and in this society.

The *ASJA Newsletter* of December 1976—the year Mort Weisinger was president—includes an exchange of correspondence between him and the Coalition Against Censorship (then a budding spin-off of the American Civil Liberties Union). "The issue involved," wrote the coalition, "is enormously important, not least because Frome's position was delicate and courageous. We would be pleased to have ASJA as one of our coalition's participating organizations." Mort in his report to members noted that the letter continued in that vein. "It hints that they are interested in funding, and perhaps we could give them some funding."

I won't presume to suggest how the Professional Rights Committee should spend this five hundred dollars, but I feel obligated and privileged to transfer this gift. Brothers and sisters, it is tough enough to make a living out of writing—but tougher to say something of more than moment but of majesty or of lasting social influence.

As the editor of *Field & Stream* told Pat McGrady, when Pat, as chairman of the Professional Rights Committee, was investigating my case, "I don't mind an exposé now and then, but why did Frome insist on naming names?" By the same token it is significant that my winning article on "The Un-Greening of the National Parks" did not appear in a major consumer magazine but in a trade publication, *Travel Agent,* owned by an imaginative friend, Eric Friedheim, who initiated the idea and was willing to expose his readers in the travel industry to the hard facts of life about our national parks.

To sin by silence when they should cry out in protest makes cowards out of men. So Abraham Lincoln declared. William Lloyd Garrison said, "I am in earnest. I will not equivocate. I will not excuse. I will not retreat a single inch. And I will be heard."

Or to quote Harry S Truman: "They said I gave them hell. But I just told the truth on them—and they thought it was hell."

That to me is what it's all about, and why I feel uplifted and honored to receive the 1981 award.

Twentieth Anniversary of the Wilderness Act—Still in Pursuit of the Promised Land

Twenty-fourth annual Horace M. Albright
Conservation Lecture,
Berkeley, California,
April 26, 1984.

I first met Horace Albright in the early 1950s. "Please don't call me Mr. Albright," he corrected me after I had greeted him. "Call me Horace." I observed in due course that he was that way with everyone, commanding respect through sheer good will and his positive, optimistic vision. Still, he was awesome, a living legend, and when I think of the marvelous people I have met and known in the cause of preservation, Horace Albright stands at the top.

He was born in 1890 in Bishop, center of the Owens Valley in California, graduated from the University of California at Berkeley, and went to Washington, D.C., in 1913 to help Stephen T. Mather organize the National Park Service. When the agency was officially established in 1916, the two of them and a secretary comprised the entire Washington staff. Mather became the first director, and they built their organization. Albright succeeded Mather in 1929, as the second director, expanding the parks into a truly national system. He stayed only four years, then left for a business career, but throughout his long life he continued to advocate preservation in public policy. For many years he was the close adviser to John D. Rockefeller Jr. and Laurance Rockefeller on major bequests and purchases they made to benefit the national parks. In the best-known example, he enlisted Rockefeller's financial support to purchase enough private land in Jackson Hole, Wyoming, to ensure establishment of Grand Teton National Park, but Rockefeller gifts, which he encouraged, were also vital to establishment of Great Smoky Mountains and Virgin Islands national parks.

When I wrote my column, "Environmental Trails," in the *Los Angeles Times* in the 1970s and 1980s, he would clip and send it to me every week, with a note or letter, often expressing his viewpoint that good people by their nature ultimately prevail, providing they do not grow overzealous or restrictive. He was a Republican, moderate to conservative, but tolerant of people like me. In a letter of June 1, 1984, he wrote, "Since you were in California, we have lost Ansel Adams, as you surely have been advised. He was eighty-two. I knew him from about 1925; was at his wedding in 1928. He was one of the greatest of photographer-artists, but he was erratic at times in his positions on conservation. I was not at all pleased with his tirade against Ronald Reagan. It was uncalled for."

I saw Albright last in 1985 at the home of his daughter in Los Angeles and in the nearby retirement home where he lived. I listened to him bring the past to life as though it were yesterday. At the retirement home he showed me that he was typing letters to Park Service widows on his trusty old portable—they were all part of the family he and Mather had started. He died two years later at age ninety-seven.

The Horace Albright Lectureship was established at his alma mater, Berkeley, in 1961, with a permanent endowment provided by contributions from friends. He delivered the first lecture on "Great American Conservationists," leaving himself out, though he certainly was one of them.

■　■　■　■

I feel highly honored at being here at the University of California, invited to deliver the 1984 Horace M. Albright Lecture on Conservation. I am blessed at the opportunity to express my ideas freely and fully before this audience in this setting, under the banner, as it were, of a man I have known and admired for years, and am privileged to call my friend, Horace Albright, one of the principal figures in the history of conservation.

Freedom of expression is paramount in my life. I say that as a journalist, but I believe that free expression is the keystone of the health and efficiency of any institution or government or society. Diversity of opinion, even dissent, challenges an institution, or a political, social and economic system, to continually review and renew itself.

As a journalist, I believe that truth telling is essential to my profession. Truth telling must and will prevail. "Knowledge will forever govern ignorance," wrote James Madison, "and a people who mean to be their own governors must arm themselves with the powers that knowledge gives." What greater goal could a journalist set for himself? What finer reputation could he earn than as one who arms the people with the power that knowledge gives?

The same is true of anyone, for that matter, anyone with knowledge and position from which to communicate it. At times, to be sure, an open expression of ideas may seem foolhardy or risky. It endangers professional acceptance and advancement. But freedom of the individual, with the right of self-expression, is sacred. I consider my freedom as a need, like water or food, to sustain the spirit as well as the body; for real success or failure comes only from within and society cannot impress it from without. To quote Joseph Wood Krutch, "Only the individualist succeeds, for only self-realization is success."

Or, as a reader of *Field & Stream* wrote to me, "History books are records of events and the doings of individuals who didn't go with the flow." Let truth hang out and consequences follow. The challenge is to make the most of the democratic American system. It may not be so good, as they say, except when compared with the alternatives. From my own life I know that it works. That I should have uncensored outlets open to my writing, that I should have a place to lecture at the University of Idaho (as I did earlier at the University of Vermont), and that I should be here tonight—such experiences give me faith in myself and the American system.

The only trouble with democracy is that we take it for granted. Democracy is what we make of it, a system under which we the people get what we deserve. Laws and regulations have their place, but only people make things work. That is why I feel that writers, and educators, too, should be leaders in the exercise of free expression. We are the human machinery that stimulates and sustains the democratic system.

Wilderness I see as the embodiment of freedom, which is why I've chosen to celebrate the twentieth anniversary of the Wilderness Act with you here tonight. That law is an extension of the charter handed down by the founding fathers with its guarantee of life, liberty, and the pursuit of happiness. Wilderness I equate with freedom from want, war, and racial prejudice, and the freedom to cultivate one's thoughts in one's own way.

Last summer while in northern Minnesota, I got to thinking about Arthur Carhart, one of the wilderness pioneers. During the period he worked for the Forest Service as a landscape architect, from 1919 to 1923, he was dispatched to the Superior National Forest, in Minnesota, with directions to prepare a plan for recreation development. Carhart, however, recognized that the area could be "as priceless as Yellowstone, Yosemite, or the Grand Canyon—if it remained a water-trail wilderness." His bosses thought that was wild talk; they were considering a master plan to build roads to reach every lake and to line the shores with thousands of summer homes. But Carhart persisted to advocate his own concept, won support, and laid the basis for establishment of what we now call the Boundary Waters Canoe Area.

Shortly before Carhart left, Sigurd Olson arrived on the scene. Over the years Olson would stand in meeting halls urging that natural values be protected from assorted mining, dam-building, logging, and motorboating. It wasn't easy and sometimes he was treated to hoots of scorn and derision. Years later Carhart paid tribute to Olson for leading a small group, which held, as he said, "a thin line of defense protecting this exquisite wilderness until help could rally to save it."

What was it they found worth defending? Based on my experience, I would call it the feel of freedom above all else. Freedom from crowds, cars, and mechanical noises. Freedom that comes from doing for one's self, without dependence on technological support. Freedom in nature, derived from being among creatures that get up and fly when they want to, or run, swim, wiggle, dive, and crawl, all admirable modes of self-propulsion. In the northern Minnesota wilderness I feel free to pick and savor wild blueberries; free to swim in cool waters, cool and dark, almost as pure as in the days of the Chippewa Indians.

I went to Minnesota as part of an exploration of wild America, pursuing adventures and encounters with different kinds of people and asking what wilderness means to them. They made beautiful statements, usually simple yet lofty and profound. One of my friends, Tom Kovalicky, a forester in Idaho, said,

> You get away from your tradition and life-style in a wilderness and you find out in a heluva hurry who you are and what you're capable of, what are the real issues in life. What really frightens you will come to the surface.
>
> Wilderness is my life-style. Wilderness is necessary. It represents that part of America that once was and always will remain. Wilderness is forever. We should be lucky enough to be smart enough to set it aside. We don't have to be like the Europeans. We don't have to

wish for that type of land representation. We'll have it.
I think we're smart in doing it.

The very idea of wilderness enriches my body, mind, and spirit, but it also elevates me to look beyond my own wants and needs. The American tradition has sought the transformation of resources; the Wilderness Act, however, stimulates a fundamental and older tradition of relationship with resources themselves. A river is accorded its right to exist because it is a river, rather than for any utilitarian service. Through appreciation of wilderness, I perceive the true role of the river, as a living symbol of all the life it sustains and nourishes, and my responsibility to it.

The Wilderness Act of 1964 opened an era of new legislation to protect rivers, trails, endangered species of plants and animals, air, water, and the environment. My travels and studies convince me that wilderness itself merits the right to be wild. Wilderness is meant for the bald eagle, condor, spotted owl, and ivory-billed woodpecker; for birds that nest in the tops of old trees or in the rotted holes in tree trunks and that need dead or dying logs to house the grubs and other insects on which they feed. Wilderness is for grizzly bears, mountain lions, big horn sheep, elk, and wolves that need large areas set aside from civilization.

We are fortunate, in America, as my forester friend said, that we have such places at this advanced stage of history. As to why our generation benefits from this legacy, I identify two principle influences.

The first of these is the influence of leadership, sometimes idealistic, sometimes practical, conceiving wilderness as a valuable entity, or resource, defining its place and purpose in national destiny, demonstrating the means of protection and perpetuation. This leadership is as old as the Republic, manifest in earlier days through the works of James Fenimore Cooper, George Catlin, and George Perkins Marsh, and in pre–World War II years of Arthur Carhart, Aldo Leopold, and Bob Marshall. The last three were Forest Service employees, which shows that government officials can be wilderness leaders, too—though few, if any at all, have made much of a mark in the last thirty or forty years.

The identification of Horace Albright with this notable lecture series leads me to discuss the early role of his agency, the National Park Service, and its approach to wilderness. From 1872, when Yellowstone National Park was established (even before there was a National Park Service), the Department of the Interior administered wilderness as a deliberate mission, even to calling troops of cavalry to protect it when Congress failed to allocate funds through nonmilitary channels. Early directors of the Park Service were strong wilderness advocates and activists. I love the story of how Stephen T. Mather, the first director, issued an order to a lumber company to dismantle its mill and depart the bounds of Glacier National Park. When the order was disregarded, Mather personally headed a brigade that exploded the mill with thirteen charges of TNT. On another occasion, when it was suggested to him that park superintendents be appointed under the same political terms as postmasters, Mather replied that he was going to pick his own people according to capability alone.

Those early years of the National Park Service were its heyday, when Mather and Albright built a vigorous, capable, aggressive organization, devoted to public service and committed to wilderness. It not only safeguarded wilderness values where they existed, but often restored these values, despite powerful opposition, in areas like Big Bend, Sequoia, Yosemite, Glacier, and the Great Smoky Mountains, all of which had been degraded by logging, grazing, mining, or settlement.

When I was writing my book about the Great Smokies, *Strangers in High Places,* Horace was a major source of historic information. He told me, among other things, about his confrontation with Senator Kenneth D. McKellar. McKellar was a rough, tough, crusty machine politician who built his strength in Tennessee by "bringing home the bacon" from Washington and by demanding every bit of his share in control of patronage (which earned him the title of "grand-pappy of all political pie-hunters"). He built his strength in the Senate through the system of seniority, which recognizes and rewards the talent of surviving through one election after another.

In the early 1930s, when McKellar learned there was to be a scenic road built on the crest of the Blue Ridge Mountains of the new Shenandoah National Park in Virginia, he demanded one just like it for the Great Smokies. The national parks director, Albright, went to see him and endeavored to explain that because of the rugged topography there could never be a road along the eastern crest of Smoky, and moreover that there never should be—that a large portion of the Great Smokies should be preserved as a roadless wilderness. The two national parks were designed to complement each other, Albright emphasized; it would be ridiculous to develop them exactly alike.

McKellar exploded. He couldn't bother with such details as the meaning of a national park and the methods of its management. He blasted Albright and for good measure blasted Albright's parentage. That a damn career bureaucrat would dare stand in the way of the welfare of Tennessee! McKellar would have liked to punch the rascal (who had to be a Republican anyway, since he was serving under Hoover) squarely in the nose, and he said so.

Albright sensed the time to retreat and withdrew. In a few days he returned to Capitol Hill. McKellar refused to see him, but Albright had brought along an intermediary, a mutual friend, who insisted that he listen and held him strongly by the arm while he did. "I will explain to you the difference between the two parks, of Virginia and North Carolina–Tennessee," began Albright. After so doing, he concluded, "I will not, under any circumstance, go ahead with the road you demand. Furthermore, Senator McKellar, I resent your personal insults."

Despite McKellar's power and influence, that road was not built. The politicians were obliged to accept the National Park Service as a bureau that ran its affairs on nonpartisan integrity. They may not have liked it, but they respected the service all the more because it lay beyond the spoils system.

Alas and alack, they don't much make them like Mather and Albright anymore. Nor like Gifford Pinchot or Ferdinand Silcox. But they do make them like John Muir; maybe not quite like the singular Muir, but in his image. This leads me to the second significant influence in the protection of wilderness.

One lesson I learned from Horace years ago is that the act of setting up national parks is not enough to make them work. By the same token, having eighty million acres designated in the National Wilderness Preservation System doesn't ensure their sanctity. National parks, national forests, national wildlife refuges, state parks and forests, county and city parks—no tract of public land has its future assured simply with a label, nor because it has a staff of paid professionals in charge. As I said earlier, laws and regulations have their place, but only people make things work. What is most needed, as Horace expressed it years ago, is "wider support from more citizens who will take the trouble to inform themselves of new needs and weak spots in our conservation program."

The fact is that each national park, starting with Yellowstone, came into being through public will and desire. Someone had a dream, plus the determination to rally others to make that dream come true. Such is the story of the Redwoods, Santa Monica Mountains, North Cascades, Rocky Mountain, Glacier, Great Smoky Mountains, all of them. The same principle applies to the ongoing protection of preserves once they are established, that is, through the identification of those "new needs and weak spots."

An individual may be struck with a brilliant idea in land use, but that idea reaches fulfillment because the people want it to. For example, the Appalachian Trail was conceived in 1921 by Benton MacKaye, a trained forester and regional planner. It was based on his wanderings in the New England forests, although others had already begun localized trails. In an article titled "An Appalachian Trail—A Project in Regional Planning," he envisioned "a 'long trail' over the full length of the Appalachian skyline from the highest peak in the North to the highest peak in the South." Few proposals in regional planning have ever fired the imagination as did MacKaye's. Scattered groups and individuals began to work, ultimately to work together to forge the Appalachian Trail into the longest marked path in the world. It has been supported ever since as something more than merely a recreational footway. "This is to be a connected trail," as the Appalachian Trail Conference declared in its constitution as early as 1925, "running as far as practicable over the summits of the mountains and through the wildlands of the Atlantic seaboard and adjoining states, from Maine to Georgia, to be supplemented by a system of primitive camps at proper intervals, so as to render accessible for tramping, camping, and other forms of primitive travel and living, the said mountains and wildlands, as a means for conserving and developing, within this region, the primeval environment as a natural resource."

The establishment of the Idaho Primitive Area is another case in point. I recently came across historic data, including the minutes of a meeting conducted at Boise in December 1930. The governor of Idaho, H. Clarence Baldridge, had appointed a committee to consider the wisdom of setting aside something to be called "primitive area" in the heart of the national forests of the state. The governor at that 1930 meeting said it was the wildest country he had ever seen and that the general consensus was it should be "perpetuated as nearly in its natural state as possible for future generations." Following considerable study and consultation, the Forest Service set aside approximately one million acres.

The Idaho Primitive Area came into existence because the people of the region wanted it to. That was in 1931, more than half a century ago. It was established on paper but endures down to our time in fact. It is sometimes argued that wilderness is the playground of elite and effete urbanites, but I don't believe it. The Idaho Primitive Area would have been lost a long time ago if the people of the state had not felt their stake in it as some priceless possession. Little wonder that, when then-senator Frank Church conducted hearings on the proposed permanence and enlargement of the Idaho Primitive Area, people who had never spoken publicly before stood and opened their hearts in praise of an area larger and wilder than Yellowstone. Little wonder that, in seeking for some appropriate way of paying tribute to Senator Church before his death, Idaho should unite in 1984 behind the new name of the Frank Church–River of No Return Wilderness, the largest wilderness outside of Alaska.

One further illustration. Twenty years ago, I attended a symposium conducted at the South Rim of the Grand Canyon on the question of whether the Colorado River flowing far below us should be dammed. David Brower, the executive director of the Sierra Club (who is here in the audience for this lecture), argued fervently that it should not be. My friend Martin Litton was present there, too, and was very active in opposing the plans of the Bureau of Reclamation, but Brower has been most identified with leadership in that issue.

I recall how Dave Brower charged at that symposium that the Sierra Club book on the Grand Canyon was being suppressed and was not available for sale in the national park. As it happened, the park superintendent of that time was a friend of mine. I was with him only an hour after Brower had made his remark. The superintendent's feathers were ruffled. "Why, of course we have the book for sale. It's right here." And there it was, hidden under the counter. The secretary of the interior, Stewart L. Udall, was one of the conservation heroes of the period, but nowhere near infallible. He was a principal advocate of the proposed dams on the Colorado River and of environmentally destructive power development in the Southwest. As a consequence, national park people were silent and silenced.

In 1970 four seasonal employees resigned their positions at Mesa Verde National Park after being warned not to discuss with visitors effects of the nearby massive Black Mesa strip-mining project on the Hopi Indians. "Morally," they declared, "we felt we could no longer work for an agency whose purpose is to protect our cultural heritage, but whose practice is censorship of major environmental problems which ultimately affect the very park in which we were working."

Those four employees should have been praised rather than forced to resign. Professional training tends to teach one to conform and direct ambitions into safe channels, whereas freedom of expression needs to be recognized, stimulated, and defended as an essential element of good government. But where government leadership fails, in the absence of a Mather or Albright, then the public voice must be heard. With due credit to the instrumental role of David Brower, it wasn't quite he who saved the Grand Canyon, but people all across this country who expressed themselves, echoing the plea of President Theodore Roosevelt early in this century: "Do nothing to mar its grandeur."

So it is with wilderness, the natural treasure that enriches our lives and our land. The Wilderness Act of 1964 could never have been passed without broad public support and approval. Thus we celebrate this twentieth anniversary of a momentous and proud happening, one of the noble achievements of modern civilization, a show of ethics and idealism to contrast with supertechnology, supercolonialism, and violence. The National Wilderness Preservation System provides hope for a coming age of reason and nonviolence, in which respect for the earth and of all its occupants will prevail.

But we have a long road to travel to realize the promise of the promised land. And I don't lay the blame on commercial interests which may or may not view wilderness as a source of raw materials. The four federal agencies responsible for administering the Wilderness System have not met their responsibility or opportunity. They don't think or plan in ecosystems. They don't direct serious attention to wilderness administration. They don't coordinate their approaches. In my travels across the country I haven't seen a single wilderness managed as it should be in fulfillment of the letter and spirit of the Wilderness Act, but I have seen wilderness areas in terrible condition, abused and degraded.

Little attention is directed to wilderness theory and principle. I've met more than a few personnel at all levels in these agencies who have never taken the time to read the Wilderness Act and consider it to be all a bother anyway. "Well," they demand to know, "how much wilderness do you actually need?" While recognizing that it can't all be wild, I feel reluctant to answer that question; what counts more is whether each succeeding generation must settle for an increasingly degraded world, reflected in degraded, circumscribed living. I can't juxtapose resource commodities against wilderness when the great value of wild country lies in its freedom, challenge, and inspiration.

We need to safeguard the sources of freedom, challenge, and inspiration. The Constitution is recognized as a sacred document guaranteeing freedom of expression, though it requires continual testing and defending. Wilderness is equally sacred, in my view—a living document of land and people, as valid and vital as the Constitution.

I've learned that we, even the experts—or especially the experts—understand very little about wilderness reserves: of how to manage and interpret wilderness so it will always be wild, of its abundant benefits to society, of how to apply the lessons of wilderness to make the whole earth a better place to live. We need to assess actual and potential values of wilderness reserves in terms of their bearing on human sensitivity and creativity.

I respect resource management and the education of resource professionals, but I want that professionalism firmly and broadly grounded, in philosophy and ethics as well as science. That's what makes being part of a natural resource learning institution exciting to me. Last year, under the auspices of the Wilderness Research Center of the University of Idaho, my colleagues and I conducted the First National Wilderness Management Workshop. The chief of the Forest Service, Max Peterson, one of the principal participants, felt moved to call it a "landmark conference," and the immediate result was to raise the visibility of wilderness in resource management on federal lands. More people now know there are problems that must be addressed. Other such projects are planned at the University of Idaho, along with ongoing courses in wilderness conservation and management.

Last month I was privileged to speak at Colorado State University on the "Twentieth Anniversary of the Wilderness Act: Heroes Who Made It Happen—New Heroes in the Making." A few years ago that presentation would have been unlikely; forestry schools didn't care much for the subject. Now Colorado State is preparing for a major national conference on wilderness research and, even more significant, to serve as host for a world wilderness congress in 1987.

These activities by educational institutions are timely. The twentieth anniversary of the Wilderness Act finds us on an upward curve of environmental concern, with wilderness at the heart of our environment, as at the heart of the nation.

I think of the campaigners for the Wilderness Act as true patriots: Howard Zahniser, David Brower, William O. Douglas, Richard Neuberger, Olaus Murie, Stewart Brandborg, Hubert Humphrey, John P. Saylor, and Sigurd Olson. My research shows Horace Albright as one of the early proponents of a Wilderness Act. At the Mid-Century Conference on

Resources for the Future, held in Washington in December 1953, he declared as follows:

> The wilderness areas in the national forests have never had a basis in law. They have been set aside by the Secretary of Agriculture for a good many years, and I am not sure that even now he can go to Congress and get such a law. Right away it would be asked: Would that mean the stopping of grazing, or of mining, or of cutting? But, just the same, law is the only means by which the areas can eventually be protected. The wilderness areas, or some of them and some of their characteristics, ought to be embodied in the law.

My favorite heroes are my own breed, writers who are activists, like Sig Olson, Dick Neuberger, Wallace Stegner, Paul Brooks, Bernard DeVoto, and especially journalists who tell it like it is, like John Oakes. In the *New York Times* of May 15, 1956, he reported that Senator Humphrey was sponsoring a bill that would set up a national wilderness preservation system. "The idea is certainly worth exploring," John wrote, "if what is left of our country in a natural state is worth saving, as many of us believe it is." He outlined the problem as follows: "This isn't just a question of city folks seeking outdoor recreation, or enjoying spectacular scenery, or breathing unpoisoned air. It goes much deeper; it springs from the inextricable relationship of man with nature, a relationship that even the most insensitive and complex civilization can never dissipate. Man needs nature; he may within limits control it, but to destroy it is to begin the destruction of man himself. We cannot live on a sterile planet, nor would we want to."

Of course we can't. The twentieth anniversary of the Wilderness Act gives assurance that we will not have to. We have still to achieve the promised land, but the pursuit itself is uplifting and yields its own reward.

Wildlife and Man in the Adirondacks: Past, Present, and Future

At a conference commemorating the one-hundredth anniversary of the Adirondack Forest Preserve, Lake George, New York, September 5, 1985.

Wherever I was invited to speak, I always felt I should "tell it like it is." It wasn't always easy, and I was never sure how the message would fly, but if my hosts didn't like it, they had the option of not inviting me back. Sometimes, though, I was surprised and felt I was tapping into feelings ready to be released.

After my talk in the Adirondacks, I received a warm letter from Rainier Brocke, a wildlife professor at Syracuse University who had organized the program. He wrote:

> We wildlifers go to our conferences, often dominated by narrow minded state game divisions, covering the same old ground and dishing out the same old hash. We talk to ourselves and piously ask why the public has not been enlightened.
>
> It has been my experience that the public includes many intelligent people in all walks of life who can put two and two together. All these vacillations, lack of leadership, lack of solid science, are not lost on them.

But I would rather look at it in a positive way: that good will come when concerned public and professionals work together.

■　■　■　■

When I was growing up in New York City, the borough of the Bronx to be specific, and dreaming of distant places, the Adirondacks could have been as far removed as Idaho in the far Northwest, while Idaho, where I live currently, could have been on the other side of the Moon. Today, the Adirondacks are only an expressway drive away from New York City, Idaho within a day by plane from almost anywhere, and the other side of the Moon doesn't seem too far either.

I did know wild animals, in the New York Zoological Park, the Bronx Zoo, fantasizing that one day I would get to observe those species in true jungle settings. Fantasy ultimately became reality, but when I arrived in Africa, South America, Alaska, and Southwest Asia the wildness of early dreams was markedly diminished by roads, dams, logging, oil drilling, factory farming, and the "golden hordes" come to observe wildlife through a picture window, with drink in hand or from a high-powered Land Rover.

In the span of a generation the world has turned upside down. Rural or agrarian culture has yielded to the industrial, the age of electricity to the age of electronics, (or whatever it may be now). When I was a kid, listening to the radio was a treat, to many Americans still then a novelty. So were the automobiles. I'm not exactly a graybeard, but I remember when riding the subway cost only a nickel. So did a hot dog at Yankee Stadium. For that matter, Babe Ruth, king of swat, in his best year received the staggering sum of eighty thousand dollars.

Where have we come from and where are we headed? In a time when one parcel of open space after another seems to vanish overnight, urbanized to oblivion and succeeded by pyramiding humanity, what is the rightful role of wild places like the Adirondacks—that is, if they have a rightful role at all? What of the animals? Are they to be found in the onrushing future only in zoos and picture books? If they are to remain as natives in the shrinking shreds of native lands remaining to them, who will save that habitat and the wild species? I feel privileged and proud to keynote the discussion of such questions at this conference commemorating the one-hundredth anniversary of the Adirondack Forest Preserve. For one thing, the centennial is an absolutely marvelous landmark of our civilization.

The 1885 action of the legislature becomes additionally impressive as the foundation for consequential actions, such as designation in 1892 of the Adirondack Park, embracing both private and public land, more than twice the size of the preserve, the largest area of the kind in the world, and the constitutional amendment of 1894, declaring that "the lands of the state, now owned or hereafter acquired, constituting the forest preserve as now fixed by law, shall be forever kept as wild forest lands. They shall not be leased, sold, or exchanged or be taken by any corporation, public or private, nor shall the timber thereon be sold, removed, or destroyed."

The authors of that amendment could have been drafting the Wilderness Act, adopted by Congress a full seventy years later. Then there was the 1970 report of Governor Nelson Rockefeller's Study Commission on the Future of the Adirondacks recommending a plan to restrict uses of private lands in order to protect scenic, recreational, wilderness, and watershed values. The commission went much further, proposing that almost one million acres in fifteen separate tracts be designated as wilderness and administered in accordance with principles patterned after the National Wilderness Preservation System. It also proposed a wild, scenic, and recreational rivers system; propagation of rare indigenous species of wildlife, such as marten, lynx, loon, and raven; and reintroduction of extirpated native species, which might well include moose, panther, and wolf. Up to that point no federal program had gone so far.

"The Adirondacks are preserved forever." So Governor Rockefeller declared on May 22, 1973, in signing legislation placing 3.7 million acres of private property within the bounds of the Adirondack Park under land-use restraints. Together with 2.3 million acres already owned and protected by the state, this entire composition of valleys, lakes, rivers, and mountain peaks became the largest area in the country under comprehensive land-use control, including wilderness designation. [This was still true early in the twenty-first century, despite many efforts to weaken protective covenants. New York state government, under Republican and Democratic control, has consistently supported the preserve.]

As history demonstrates, however, nothing is ever "preserved forever." I'm not even sure that it should be, and at times I've been called an extreme preservationist. Once-treasured places may be lost forever, but once saved they need to be resaved over and over, again and again, their propriety reviewed, defined, defended, and renewed in a changing society. That is, if society chooses to save and renew them. Old or new laws and regulations may have their place, like government itself, but only people make things work.

This conference is a manifestation of our democracy, with its broad representation of participants, professional biologists, public officials, conservation leaders, and private citizens here to consider a serious and extensive program. While focused specifically on the Adirondacks, we are weighing principles of wilderness and wildlife that may, and hopefully should, be applied universally. Let the Adirondacks, as in times past, show the way for times future!

My research and observations tell me that Americans want and support their wild places, that they want them protected and preserved as part of a healthy, wholesome society. I could say that the advance of technological supercivilization makes the remaining fragments of original America more priceless, but that would be a subjective value judgment on my part. Nevertheless, I believe the little people are way ahead of their elected officials, public administrators, resource professionals, and the institutions of learning where these professionals are trained. Enrollment at most forestry and natural resource schools has been steadily declining, not simply because "the jobs aren't there any longer"—there's plenty of work and challenge for those concerned with sustaining and restoring land health—but because a new generation demands a new curriculum to replace the old rules and roles predicated on commodity production. The new curriculum must replace the old, designed as it was for employment in exploitive industries and in tired, inbred federal and state resource bureaucracies. [A number of these institutions have changed names from "forestry school" to "school of natural resources," not always, however, with new curriculum.]

Evidence that Americans support wilderness is in the opinion polls underwritten by the timber industry that produced answers the sponsors never expected. I see further evidence in statements of elected public officials and candidates for office. None speaks for elimination of wilderness. Even an ardent public-lands foe like Steve Symms, the junior senator from Idaho, is likely to go before the electorate to say, "We have enough wilderness already" or "We must balance preservation with jobs" rather than "We should not have any wilderness at all."

I see evidence in the record of the Reagan administration, which came on strong in the first term, determined to undo wildland protection. However, President Reagan and company learned they had mistaken their mandate. Before the election for a second term the Sagebrush Rebellion, the proposed giveaway of western public lands, was withdrawn. James G. Watt was dismissed as secretary of the interior; before the election it was clear that John B. Crowell, the other administration anti-wilderness spearhead, was on the way out as assistant secretary of agriculture, where he had been ruling and ruining national forests. And Anne Gorsuch Burford was likewise dispatched from the Environmental Protection Agency.

The president in 1984 signed into law legislation adding eight million acres to the National Wilderness Preservation System. He chose not to make an event of it, yet he officiated at adding the greatest amount of land to the Wilderness System since passage of the Wilderness Act twenty years before—that is, except for the huge Alaska acreage of the Carter administration. He was wise enough to avoid exercising a veto. President Reagan may have his private feelings, but his administration has definitely softened, and may even have improved somewhat, at least in its projected image.

The most singular evidence of public support is in the land itself. Laws establish wilderness only on paper. Yet here in the Adirondacks, Idaho (enriched by more classified wilderness than any state except Alaska), Alaska, and throughout the country, wilderness exists in fact. It has endured, if you ask me, despite resentment from industries and resistance from professional resource managers, because people of all walks of life want it, and feel the need of it, as part of our modern civilization, a show of ethics and idealism to contrast with the supertechnology, supercolonialism, greed, and violence that seem characteristic of our age.

Wilderness provides a wonderful resource for learning, learning in many fields, including wildlife science, human behavior and ethics, all related. Aldo Leopold in a 1941 essay on "Wilderness as a Land Laboratory" expressed it very well:

> Paleontology offers abundant evidence that wilderness maintained itself for immensely long periods; that its component species were rarely lost, neither did they get out of hand; that weather and water built soil as fast or faster than it was carried away. Wilderness, then, assumes unexpected importance as a land-laboratory. . . .
>
> All wilderness areas, no matter how small or imperfect, have a large value to land science. The important thing is to realize that recreation is not their only or even their principal utility. In fact, the boundary between

recreation and science, like the boundaries between
park and forest, animal and plant, tame and wild, exists
only in the imperfections of the human mind.

Our generation is still catching up with the full meaning of this statement. In autumn 1983, colleagues and I at the University of Idaho conducted a national wilderness management workshop. It was well
attended and thoroughly stimulating, a worthy exercise in learning. We
have since been endeavoring to disseminate the lessons, all of which
derive somehow or other from the Leopold reference. I would summarize these lessons as I perceive them from that workshop and other
experiences as follows:

- The earth is here for us humans, but not for us alone. We may
  perch atop the pyramid of life, but the fact of superiority and
  power imposes special responsibilities as well as rights. We exist
  and enjoy the world only by virtue of conditions created long
  before the arrival of humankind and through the millennia by all
  other forms of life. The earth's crust is a living organism, and all
  life an extension of ourselves.

- The manipulation of food and cover is based on the traditional
  anthropocentric philosophy that wildlife and fish must be used
  like a commodity to be considered useful. Manipulation is one
  approach, wholly focused on "production" of desirable game
  species. Abundance of game is critical to state fish and game
  agencies because of their dependence on sale of licenses and on
  equipment taxes for their operating revenues. But it isn't enough
  to know there are millions of deer, elk, wild turkey, pronghorn,
  pheasant, bobwhite quail, and mourning dove "that can survive in
  an environment molded by human beings," as some game biologists like to put it, for the idea of preserving wildlife gains legiti-
  macy when applied to our relationship to the total environment
  and not to isolated parts of it.

- In wilderness at times there is apt to be less game because of the
  slow, inevitable natural process of supply and demand to sustain
  life in balance. Yet elk, grizzly, moose, caribou, mountain sheep,
  and mountain goat, sometimes considered a wilderness-dependent
  species, are normally classed as game. In addition, however,
  whatever is natural belongs, whether considered beautiful or

ugly, useful or not useful, for each organism fulfills a distinct function in its own right, whether we recognize it or not.

- The principal focus of concern thus far has been on large mammals and birds, but it's not only the grizzly, polar bear, wolf, sea turtle, whale, eagle, and falcon that need help. It's all of the greater and lesser creatures. In the last two centuries humankind has wiped out an estimated ten thousand species of insects and snails, scantly referenced in lists of endangered species, ignored in natural resource inventories, yet indispensable in the cycle of life.

- The disappearance of species reveals flaws in our social and economic system. The expansion of cities, construction of highways and power dams, the use of toxic chemicals in agriculture and industry, acid rain, dumping of heavy metals in water, destruction of forests, the increase in leisure activities sending millions of people into wildlife territory—all those impose heavy pressures. Conversely, the challenge to save wilderness species provides the chance to do something for our own species as well. I learned this anew on a recent trip to California. The California condor, once numerous and widespread, is near extinction; the known number of birds remaining in the wild is now down to eight. It's not difficult to understand why, considering the current condition of southern California, with the blight of galloping human development. The condor can still be saved, through protection of the mountain wilderness where it breeds, nests, and roosts and restrained human uses of the valleys where it forages for food—in other words, some kind of adaptation of the Adirondack Park plan. That would not only help the condor but also keep southern California habitable. [Condor numbers remained low into the twenty-first century, but the trend to extinction appeared to be reversed. Condors were released in Arizona, with considerable promise.] I learned the lesson again in Wyoming, where the 1982 count of bighorn sheep was 6,400, as compared with an estimated 150,000 or more a century ago. Despite a slow increase under protection and management in recent years, a precarious balance between birth and death rates keeps the future uncertain. "A very slight increase in natural mortality," warns the Wyoming Game and Fish Department, "could 'tip the scales' in the opposite direction. This suggests we need to be extremely careful in the management of land uses and human activities that carry the potential to impact sheep populations."

■ A civilized nation ought to be able to sustain itself by devouring less of the earth's resources, or by utilizing them more efficiently, without further sacrificing the habitats of bighorns, condors, grizzly bears, wolves, panthers, or any other species. But who will lead the way? A new alliance of hunters and nonhunters, who share the same fundamental goals, can do it. I've learned that few species are threatened by the gun. True sportsmen should not be held responsible for the sins of slob hunters and wealthy dilettantes who are likely to have their guides kill their trophy animals for them. When persons of common interest allow themselves to be divided, somebody else will benefit. James G. Watt knew this when he sought to drive a wedge between hunters and the environmental community. In the wildlife crusade that I envision there is room for all—for all willing to concentrate on the priority of saving wild places for wild animals.

■ One final point on "imperfections of the human mind" to which Leopold referred. We live in a generation programmed to taking rather than giving, to success measured in terms of acquisition of material wealth rather than in caring and sharing. I recently read an interview with Adolfo Perez Esquivel, winner of the 1980 Nobel Peace Prize and Argentine human rights leader. How, he was asked, can a person live in harmony with the essence of life? To which he replied:

> When one talks of nature one does not mean simply to admire nature, but to share it, in the sense that if food is needed, one should take what nature gives us, but not with primitive blindness, pointlessly destroying nature out of greed. We need to develop respect for nature, and learn to use this wisely. . . .
>
> In the schools, children learn the history of power as the power of domination and conquest, instead of learning that power is necessary insofar as it is the power of service. If we do not have the consciousness of service to our fellow beings, power will continue to

be utilized as the power of domination. We will continue to find ourselves caught up in violence.

Thus I foresee this conference endeavoring to prescribe an antidote
to violence, to ensure the future of the lovely Adirondacks, pride of the
East, pride of all America, as the sanctuary of hope, as much for our own
kind as for the wild itself.

Could My Words Possibly Make Any Difference?

Accepting the Marjory Stoneman Douglas Award, presented by the National Parks and Conservation Association, Washington, D.C., April 7, 1986.

This was the night, as I mentioned earlier, when Marjory Stoneman Douglas stole the show. There wasn't a thing I could do about it anyway, and as my friend Stewart Brandborg summed it up at breakfast the next morning, "She's entitled—we should both do as well when we get to be ninety-six."

It was still a great event. I sat on the dais next to William Penn Mott, director of the National Park Service. I felt that was an honor in itself. Mott was a straight arrow who had been director of California state parks, probably the best state park system in the country, and now was doing his best to protect the national parks from being politicized and poisoned by Reaganauts running the Interior Department. Paul Pritchard, president of the National Parks and Conservation Association, said kind words in presenting the award, and I looked out at a friendly crowd of 150 or more public officials and environmentalists, some dear friends and others my sparring partners, all of whom cheered, then cheered again when I said I was giving away the award money.

■ ■ ■ ■

I feel deeply grateful, humbled, inspired, and, above all, hopeful that I will prove worthy of the trust and confidence implicit in this award. I'm reminded of attending a reception several years ago for Ansel Adams when he was about to receive one of the many honors that came to him. I told Ansel that it was wonderful his creative genius should be recognized and rewarded. He was feeling good about it, yet on the modest side. "My mother told me," Ansel reflected, "that if I just kept at it, I might amount to something."

Gifford Pinchot said it a little differently: "The most powerful thing in human affairs is continuity of purpose."

For myself, I feel privileged to have spent my time and energy in behalf of conservation and national parks. Over the years many friends have encouraged me. A number of them are here tonight. I can't mention names, but those in particular who started me on the path more than thirty years ago were Sam P. Weems, Herbert Evison, Ronald Lee, and Conrad Wirth of the National Park Service and Clinton Davis and Richard McArdle of the Forest Service. I wish they were here, too, but all have gone to their reward.

When I was younger, beginning my career as a journalist, I wondered whatever became of all the words I was cranking out. Did anybody ever read them? Could they possibly make any difference?

I discovered in due course that I had chosen the most wonderful way of life. While I was a columnist for *American Forests* I received a letter from a woman engaged in efforts to prevent mining in the mountains bordering her hometown, Colorado Springs. "The most beautiful word in the English language is hope," she wrote, "and you have given meaning to it for us here." A column I wrote became the rallying point for the heroic and ultimately successful crusade to save Overton Park, the beautiful forest in the heart of Memphis. Those words *did* make a difference.

During my time as conservation editor of *Field & Stream* I received many, many letters from readers, little people all across America, telling me of their special causes and concerns. "Have we so much of earth that we can afford to sacrifice any part of it?" So demanded a woman in western Montana, giving me a challenge I have never been able to put down.

While accepting the Marjory Stoneman Douglas Award with personal pride, I can't help sharing the occasion with all those who have taken responsibility for safeguarding a special fragment of earth that holds meaning to them. Every single piece of earth has its own place, personality, and purpose. Some individual comes along to show us why it is worth saving and how to save it. Years ago when I went to Florida in navigation training, the Everglades seemed to go on forever and were considered to be a wasteland. Marjory Stoneman Douglas explained that here is a precious river of grass deserving of our protection.

The same is true of every parcel that has been or ever will be preserved. Someone has the dream and the courage to make it come true.

"The only thing necessary for the triumph of evil is for good men to do nothing," Edmund Burke declared. Because I agree with Burke, I want to share the gift to me of five thousand dollars with good men and women with the vision and conviction to fight for their parcels of earth:

Idaho Conservation League, to honor Paul Fritz.

Utah Wilderness Association, to honor Dick Carter.

Montana Wilderness Association, to honor Riley and Pat McClelland.

Friends of Wyoming Wild Deserts, to honor Lynn Kinter.

Friends of the Columbia Gorge, to honor Nancy Russell.

Friends of the Santa Monica Mountains, to honor Susan Nelson.

Texas Committee on Natural Resources, to honor Ned Fritz.

Friends of the Boundary Waters Canoe Area, to honor Chuck Stoddard.

Tennessee Citizens for Wilderness Planning, to honor Mack Prichard.

Northern Virginia Conservation Council, to honor Ellen Pickering.

The generous gift of Bon Ami, the confidence of NPCA president Paul Pritchard, the presence here of Mrs. Douglas mean all the more to me. I feel proud to accept in behalf of my profession of journalism. It proves that a journalist can be involved as an individual without hiding behind the excuse of "objectivity." Models of my profession have been activists—Edward Meeman, Richard Neuberger, John Oakes, Bernard DeVoto, Sigurd Olson, and my friend and contemporary Bob Cahn, who is here tonight.

[Robert Cahn won a Pulitzer Prize for a 1968 series of articles about the national parks in the *Christian Science Monitor*. He later was appointed to the Council on Environmental Quality by President Nixon.]

I note with pride that Stephen T. Mather was a journalist on the *New York Sun*. He made money later and put it to work in causes like this. He contributed to get the National Parks Association going. One single dollar invested begets dividends. There are so few, they say. Yes, but it begins with few. So let us go on together from here.

For Love of the Beautiful Counts Most

Forum of the New York Parks and Conservation Association, Saratoga Springs, New York, March 7, 1987.

Early in 2001 I noted an item in one of the environmental journals showing once again that land designated for protection is not necessarily protected for perpetuity, not even in a seemingly progressive state like New York.

In this particular case, Eugene Meyer, best known as owner and publisher of the *Washington Post,* had acquired in his early business career in New York a two-hundred-acre suburban estate called Seven Springs Farm at Mount Kisco. He willed the property on his death in 1969 to the Eugene and Agnes Meyer Foundation, a nonprofit corporation, and provided money to purchase adjacent Byram Lake from the city of New York, with the explicit proviso the lake be utilized in the public interest for water supply or as a park.

The foundation in time decided to donate the property to Yale and Rockefeller Universities, which converted it into cash for their coffers, seven million dollars' worth, by selling it to Donald Trump, the New York developer. He announced plans to profit on his investment by building a first-class golf course. Local critics insisted the golf course would contaminate Byram Lake with pesticides applied to keep the greens spotless and free of grubs and siltation from construction, and that it would disturb an adjacent Nature Conservancy preserve. As I declared at Saratoga Springs, places of scenic beauty are being reduced and degraded. Our responsibility is to see that future generations enjoy the same opportunities for solitude and the same sense that nature prevails that we have known.

■ ■ ■ ■

I've talked to groups of concerned activists in various parts of the country about "citizen leadership" in the protection of parkland and open space. Here and there people ask how I got started in my career in conservation and where I came from, assuming that I was raised in someplace like Montana, or at least in Minnesota or Missouri, and that that would have something to do with my interest and activity. When I explain that I was born in Brooklyn and grew up in the Bronx, in the city-most of all cities, they respond with surprise as though city people have no basis for connecting with nature.

Coming home to New York, at least to New York State, gives me the opportunity to look back and reflect and hopefully to connect with the subject at hand. Yes, I acknowledge that kids in my time played stickball and roller hockey in the streets and softball and basketball in the school-yards. But I remember the nice parks we had, green places that broke the concrete canyons—including Van Cortland Park next to the high school I attended at the north end of the Bronx, and the park sur-rounding the cottage where Edgar Allan Poe once lived, with benches and shade trees on the Grand Concourse.

I had an uncle who regularly released pigeons from a coop on the roof of an apartment house in upper Manhattan, and I thought that was a special exodus from the urban—for the pigeons and for him. I read John Kieran, sports columnist in the *New York Times* and loved his descriptions of birding in Central Park in the heart of the city. He taught that nature was in the air and under foot.

I could ride the subway for a nickel to Coney Island or Brighton Beach and swim in the Atlantic surf. Or I could and many times did walk from the family apartment to the Yankee Stadium, buy a ticket in the bleachers for fifty cents and a hotdog for a dime while watching Babe Ruth, Lou Gehrig, Tony Lazzeri, Red Rolfe, Red Ruffing, and Bill Dickey, or cross the bridge to the Polo Grounds to watch my heroes of the Giants, Bill Terry, Carl Hubbell, Hal Schumacher, Travis Jackson, Shanty Hogan, and Mel Ott, in the Polo Grounds. Those teams played on real grass and under natural sunlight, and starting pitchers stayed in the game for nine innings.

The whole world has turned upside down. Baseball players earn a million dollars or more; the eighty thousand dollars Babe Ruth earned at the peak of his career is a pittance, maybe a month's pay for a rookie. Open space, elbowroom, roving room, however you choose to call it, has drastically diminished. At the turn of the twentieth century John Muir commented on the abundance of wilderness still evident across the continent, but this is no longer the case. Whatever wilderness still remains is vulnerable to human intervention. Anywhere I go, whether the Adirondacks in New York or the Sawtooth Mountains in Idaho, I can be sure to meet high-tech hikers, backpackers, and climbers, the "gearheads," "techweenies," and "Pataguccis."

Something else has changed. When I consider natural resource history, I think of strong-willed leaders in both federal and state governments. Gifford Pinchot, first chief of the U.S. Forest Service, and Stephen T. Mather, first director of the National Park Service, and others were crusaders at heart. Pinchot, after leaving Washington in 1910, returned home to Pennsylvania, where he was twice governor and pressed for establishment of state parks and forests. Mather believed that success of the national parks depended on a connecting network of state parks and campaigned for them. In his home state, California, Mather in 1919 helped organize the Save the Redwoods League, which has since put together thirty redwood state parks comprising 135,000 acres.

Oswald West, governor of Oregon from 1911 to 1915, kept one of the most beautiful seacoasts on earth from being exploited for private greed and gain. Percival Baxter, governor of Maine from 1921 to 1925, failed to convince the legislature to protect Mount Katahdin, the state's greatest mountain, but that did not stop him. Determined to act as a private citizen, Baxter in 1930 with his own money purchased six thousand acres, including most of Katahdin, which he deeded to the state. He continued to buy land, increasing the size of what is now known as Baxter State Park to two hundred thousand acres.

When Governor Franklin Roosevelt of New York became president, his national recovery program encouraged the development of state parks across the country. Many states organized park systems for the first time in order to qualify for federal funds and benefited further from projects of the Works Progress Administration and Civilian Conservation Corps. Even now in state parks of New York, Indiana, Kentucky, Washington, Oregon, California, and other states I still find structures, trails, and park designs of those days still in place and serving useful purpose.

But times have changed. State parks in too many instances have been opened to logging and hunting and highways. They have been transformed into resorts with luxury lodges and golf courses, operated by profit-making private corporations. Because legislatures have cut their budgets, park managers in some instances have been directed to raise their salaries through commercial ventures inimical with true park purpose.

The initiative for protection and preservation clearly has passed from the professionals in government to the people. I have only to look for evidence here in New York. I dread to think of what the landscape of this state would be like without citizen concern and involvement in issues of the Adirondacks, the proposed power plant at Storm King on the Hudson River, the Appalachian Trail, the ill conceived West Side Highway in New York City, and even dear Central Park.

The truth is that places of scenic beauty are not increasing in number but are being reduced and degraded. Our responsibility is to see that future generations enjoy the same opportunities for solitude and the same sense that nature prevails. Charles Evans Hughes, governor of New York in the first decade of the twentieth century, had it right at the dedication of Palisades Interstate Park when he said, "Of what avail would be the benefits of gainful occupation, what would be the promise of prosperous communities, with wealth of products and freedom of exchange, were it not for opportunities to cultivate the love of the beautiful? The preservation of the scenery of the Hudson is the highest duty with respect to this river imposed upon those who are the trustees of its manifest benefits."

We need to believe and echo these words, and to rekindle love of the beautiful in public policy and professional performance. I have a story to tell about Sam Boardman, "father of the Oregon state park system." After retiring in 1950 at the age of seventy-six, Sam wrote the history and description of the parks until he died three years later. Reading Sam's memoir I learned about Beacon Rock, one of the scenic features of the Columbia River Gorge. It stands by itself at the edge of the river, on the Washington state side, as a massive beacon of stone. Sam recorded that in 1931 Beacon Rock was considered for rock purposes in construction of jetties at the mouth of the Columbia River. Washington state was ready to let that happen, but not Sam—he found the owner willing to donate the land to Oregon and urged that it be acquired as a state park, based on this rationale: "Why should we just let the width of a river destroy a scenic asset woven into a recreational garland belonging to both states? How could we stand by and see the death of a relative, though a bit distant? If such things of beauty were not fought to a saving conclusion, then the waters of Multnomah Falls would be falling through steel pipes for the generation of electricity."

The *Portland Oregonian* reported Sam's idea and editorialized: "So far as we know no state in the union now owns a park in another state.

It is a gift unprecedented and as such, whether Oregon accepts it or not, is likely to arouse a stronger public interest in this natural monument."

So it did. Home pride, as Sam Boardman wrote, was rekindled through the effrontery of a bordering state. Washington made it a state park, and Sam wrote, "It is ours to see and wonder at its birth."

During my boyhood in the Bronx, friends and I would take the ferry from Alpine to cross the Hudson River and hike the Palisades. I never knew then how builders one hundred years ago decided to crush these towering prisms of rock, the Palisades, into concrete, or how people reacted and rallied until the Interstate Park Commission purchased and silenced the cliffside quarries. But I do know now how the majestic cliffs came to be protected as Palisades Interstate Park. It is ours to see and wonder at its birth, and ours to safeguard for the love of the beautiful.

Why Must We Save the Forest from Foresters?

Annual pow-wow, Texas Committee on
Natural Resources, Indian Mounds Wilderness,
Texas, April 4, 1987.

One of my manuscript reviewers noted the absence of an introduction to this speech and suggested I ought to include one to be consistent. So I will set down a few thoughts about woodlands in the South, which to my mind extend roughly from Virginia or southern Maryland in the East to Arkansas, Oklahoma, and Texas in the Southwest.

Once, in the 1960s, the Forest Service engaged me to write a booklet, "Trees of the Forest—Their Beauty and Use." I became taken with the beauty aspect but was reminded to emphasize the uses as well, especially of the commercially valuable pines of the South. Maybe so, but the botany and biology of these forests compare with those of the Amazon rain forest of South America. Certainly William Bartram was overwhelmed with the variety and beauty of what he discovered in his classic eighteenth-century exploration, and he never made it as far as the bayous of Louisiana or the Big Thicket of Texas.

I made speeches in different parts of the South and never said that trees should not be harvested appropriately for use. But that southern province has been exploited ruthlessly by esteemed corporations that took what they wanted, then left communities destitute and moved on. Colleges and schools ought to teach more of that history.

■　■　■　■

Being out there today in the lovely Indian Mounds Wilderness, celebrating the glory of God's creation, I felt anew that the most important gift our generation can make to the future is to save some of it, to place earth first in our hearts and minds, to make preservation of wild nature a priority for society.

I ask my friends of the Forest Service, the legally designated guardians of the Indian Mounds, to join with us in this commitment. They, like we, will be remembered not by the volume of board feet logged, the volume of chemical poured on the earth, the miles of road built, the sediment in the streams, but by their positive stewardship of God's green earth.

We are here in the domain of southern pine forests, where one-hundred-feet-tall loblolly pines grow in the good company of beech, magnolia, and white oak, and where the creek bottoms and glades are alive with wild orchids, hollies, azaleas, trillium, and ferns, and with the sounds of warblers, vireos, woodpeckers, and wrens.

Most of the southern pine forests, the piney woods of history and legend, are gone. I am reminded in this setting that forestry was born out of destructive logging, out of the damage done by timber barons, who saw only product and profit and once they reached the maximum moved on. Fierce fires followed, fed by resinous tops left on the ground. In most areas the pine forest was replaced by a jungle of scrub oak and catbriers.

Thus the national forests of Texas, the Davy Crockett, Angelina, Sabine, and Sam Houston, and national forests elsewhere in the South, were established to give nature a second chance on lands that nobody wanted. And now, after a long period of hard work in conservative forest culture by the Forest Service, the forests have grown back—although not exactly whole, I fear, for we likely will never again see the ivory-billed woodpecker, which perished with destruction of its habitat by logging.

That is all the more reason to cherish the national forests. I contrast these public natural forests with the millions of acres around them, where forests have been reduced to tree farms, huge plantations designed as cellulose factories, with every tree the same age as its neighbor and looking exactly like it. Little wonder that beetle epidemics feast on this monoculture of pine.

I came here to East Texas in April 1976 to meet Edward C. "Ned" Fritz, the founder and prime mover of the Texas Committee on Natural Resources, or TCONOR. Ned was a Dallas trial lawyer who quit his practice to work as a full-time unpaid activist. He and I went together by arrangement to inspect the proposed Four Notch/Briar Creek Wilderness southeast of Huntsville. We made the trip in response to a summons we both had received from Madeline Framson, a member of the Sierra Club in Houston, who was deeply concerned about logging of big pines and hardwoods underway within the wilderness study area. At the site, we met Madeline, and also the supervisor of the Sam Houston National Forest and the district ranger responsible for the Four Notch area.

The foresters were cordial but clearly focused on producing timber by clear-cutting—the cheapest, easiest way—and were absolutely scornful of wilderness designation and even of recreation as a valid forest use. Framson, Fritz, and I all were distressed. Ned went home, found two lawyers willing to work *pro bono* and sued the Forest Service.

I attended the trial in December 1976. The government agency clearly benefited from intervention by the timber industry represented by a battery of Houston lawyers. TCONOR for its part introduced a variety of specialists, including Gordon Robinson, consultant to the Sierra Club, formerly an industrial forester with many years experience, and Charles H. Stoddard, a retired government official and author of a textbook, *Essentials of Forestry Practice,* both of whom came on their own and worked for free. On May 24, 1977, the judge, appropriately named William Wayne Justice, issued a permanent injunction against clearcutting in national forests in Texas, due to impairment of recreation, wildlife, and soil conservation.

The government appealed and the judge's decision was overturned. But the issue did not go away and the process of democracy is still in place. People who care, who care enough, can have their way. The only thing wrong with our democracy is that we don't always utilize it, or test it to the fullest. But when we do, then look out! In the early 1970s, the Forest Service said, "There is no wilderness in the East." But there *is* designated wilderness in the East today. In fact, the regional forester of the Eastern Region, Floyd Marita, reported recently that 17 percent of national forest land in the East is now in the Wilderness System and there can be more still. I remember how an earlier regional forester in the South, Ted Schlapfer, disputed the Alabama Conservancy in its efforts to protect the Sipsey area of the Bankhead National Forest, and how his agency resisted the Conservancy's proposal for wilderness. But the citizens persisted. In 1975, the Sipsey Wilderness was established with 11,000 acres, and has since been enlarged to more than 20,000 acres.

I will repeat my words that concerned citizens and the Forest Service should be on the same side, in cooperation, not opponents in constant confrontation. The heart of the issue is sound land management.

The Forest Service policy of promoting and practicing clear-cutting, reckless and destructive as it is, must end. Clear-cutting must be replaced by conservative, responsible, respectful treatment of the land, with production of timber recognized as only one use of the multiple uses mandated by law and by ecological principles.

I find support for what I say here from the ranks of the agency and the profession. Roy Feuchter, director of recreation for the Forest Service, wrote in 1979, "Foresters are generally timber-oriented and not people-oriented. Our biggest training job is to convince the wildland managers, not the public, of the values of outdoor recreation."

And Carl Reidel, director of the environmental program and professor of forestry at the University of Vermont, in 1980: "How can foresters be trusted to manage a natural preserve when they can't identify the birds or wildflowers in the area, much less understand their special environmental requirements? How can foresters even be trusted to mark timber for intermediate cutting if they are insensitive to the practical problems of felling a tree or locating a skid trail on a rocky side hill?"

Likewise, Tom Coston, regional forester, Northern Region of the Forest Service, in 1981: "Some administrators in the past have felt public involvement was cumbersome and a stumbling block in the way of efficient management. That I don't accept. It has proven a great benefit to the national forests of the Northern Region in Montana and Idaho. Participating individuals and groups help Forest Service personnel make better decisions and confirm the democratic approach to public land management. National forests don't belong to just those working in the Forest Service—they are part of our national heritage."

All over the country citizens are concerned, distressed, and angry. It is time for the agency to wake up. Why must the people save the national forests from the foresters in charge?

[By the late 1990s, under pressure of the Endangered Species Act and other environmental laws, the volume of timber logged on the national forests was drastically reduced. Forest Service leaders shifted attention from logging to development of recreation as a major function. Much of the emphasis, however, was placed on off-road vehicle (ORV) activity and on "partnerships" with commercial entrepreneurs ready to provide services on public lands.]

William O. Douglas, associate justice of the Supreme Court, champion of wilderness and civil liberties, exemplifies twentieth-century intellectual independence and courage. In 1954, after the *Washington Post* editorialized in favor of building a motor parkway along the Chesapeake and Ohio Canal, Douglas stepped down from the bench to lead a 185-mile protest hike the full length of the canal from Cumberland, Maryland, to Washington, D.C. He insisted the towpath was a linear sanctuary "not yet marred by the roar of wheels and sound of horns." The *Post* editor joined the hike and changed his mind. In 1977, the Chesapeake and Ohio National Historical Park was officially commemorated to Justice Douglas "in grateful recognition of his long outstanding service as a prominent American conservationist and for his efforts to preserve and protect the canal and towpath." National Park Service photograph.

In American Automobile Association days of the early 1950s I went West (with colleague Jerry Fisher, *center*), here visiting Virginia City, Nevada, in company with Lucius Beebe, expatriate New York columnist and bon vivant who revived the *Territorial Enterprise* where Mark Twain got his start in journalism. On this same trip we went to Yosemite National Park, which proved more important in the long run.

Chained- and caged-animal attractions, like the bear roosts at Clark Lake, New Hampshire, shown here, and reptiles under glass at saloons, in the South and West, were meant to entice tourists in the 1950s and 1960s. I felt them repulsive and shameful to states that tolerated them. Photograph by Mack Prichard.

President Lyndon B. Johnson signed the historic Wilderness Act in the Rose Garden of the White House, September 3, 1964. Representative John P. Saylor of Pennsylvania, who introduced and fought for the Wilderness bill in the House of Representatives, stands directly behind the president, flanked by Secretary of the Interior Stewart L. Udall *(right)* and Secretary of Agriculture Orville L. Freeman *(left)*. The two women in the front row center are Margaret Murie, whose husband Olaus, a leader of the Wilderness Society, died the year before, and Alice Zahniser, whose husband Howard, the principal architect of the Wilderness Act, died only four months before this ceremony. Photograph courtesy of LBJ Memorial Library.

Defense Department news release, February 8, 1967: "Assistant Secretary of Defense Paul R. Ignatius (*fourth from the left facing*) and members of his staff met today with the Natural Resources Council of America to discuss the importance and future role of Defense in natural resource activities." Ignatius is quoted in the news release as declaring, "Conservation of our natural resources is a vital part of Defense, for without natural resources of all types, in general abundance, there can be little national security." But military agencies have transformed significant areas of the United States, Puerto Rico, and other nations into bombing ranges and mock battlefields. In this photo, I am seated at the extreme right, representing Defenders of Wildlife, with my close friend Rupert Cutler, on my left, representing the Wilderness Society (and later assistant secretary of agriculture). Defense Department photograph courtesy of Rupert Cutler.

Harvey Broome, president of the Wilderness Society *(left)*, congratulates Ernie Dickerman, his cohort and longtime hiking buddy, on winning the Chevron Conservation Award at the annual Appalachian Trail Conference, Cashiers, North Carolina, May 1967. Broome died less than a year later, in March 1968. Dickerman carried on until his death in August 1998. An obituary in the *New York Times* referred to him as "'the granddad of Eastern wilderness' for his efforts to persuade people to defend their forests and parks from power saws and bulldozers." Photograph by Mack Prichard.

Interior Secretary Walter J. Hickel (*left foreground*) on June 23, 1968, responded to an appeal by one hundred conservationists by overruling George B. Hartzog Jr., director of the National Park Service (facing Hickel), and his plans for new roads and massive campgrounds in the Great Smoky Mountains. Leroy Fox and Walter Williams (arms folded), Tennessee activists, are at left. Ernie Dickerman, who led the delegation to Washington, is second from right. Photograph by Mack Prichard.

John P. Saylor *(center)* was a two-fisted, fighting congressman from western Pennsylvania. He was my hero and I always felt honored in his presence. We are pictured here in Washington with Michael O'Neill, publisher of *Field & Stream,* in 1972. "Big John" was a Republican conservation crusader in the mold of Theodore Roosevelt and Gifford Pinchot.

This was forestry as practiced by the U.S. Forest Service on Admiralty Island, Alaska, in 1972, disrupting the habitat of Alaskan brown bear and bald eagles. I made the field trip to Admiralty in company with Mike Miller, a writer friend and state legislator, and Karl Lane, a master guide who joined the Sierra Club in a lawsuit to protect Admiralty. That year Trout Unlimited gave me its Trout Conservation Award. Photograph by Mike Miller.

As a teacher, I encouraged students to get involved in the democratic process and learn by doing outside of the classroom, with or without credit. In 1993, Western Washington University students took me seriously, joining protests against logging of the ancient wild forest at Clayoquot Sound on Vancouver Island, British Columbia. Those shown here are at the Macmillan Bloedel logging docks at Tacoma, Washington. Photograph by Michael Rayton.

Reps. Morris "Mo" Udall, Arizona *(left)*, and John Seiberling, Ohio, two environmental heroes, are shown in Fairbanks, Alaska, for hearings on H.R. 39, the Alaska National Interest Lands Act (ANILCA) in August 1977. Udall, as chairman of the House Interior Committee, appointed Seiberling chairman of a special subcommittee to deal with public lands in Alaska. Seiberling conducted hearings in major cities in the Lower Forty-eight and in Alaska, listening to testimony from more than twenty-three hundred witnesses. "In a single vote we doubled the size of the National Park System and preserved the 'crown jewels' of the continent for posterity," Udall said later, "in large measure because of the dedication of John Seiberling." Besides the large new parks, H.R. 39 established wildlife refuges, wild and scenic rivers, and major additions to the National Wilderness Preservation System. Photograph courtesy of John Seiberling.

I prayed that I was not looking at the future of the national parks, nor at the future of American taste, while stuck in this immense traffic jam, worsened by the abundant East Tennessee roadside commercial tourist attractions leading to Great Smoky Mountains National Park, on July 30, 2001. Photograph by Mack Prichard.

Chow line in the Bridger Wilderness, Wyoming, with Trail Riders in the Wilderness (trips then conducted annually by the American Forestry Association), 1960. Forest Service photograph by Clint Davis.

High in the Sawtooth Wilderness, Idaho, in 1964, with Forest Service officers Jack Lavin (*standing*) and Max Rees. They and people like them showed and taught me a great deal in the field about taxonomy, dendrology, silviculture, hydrology, geology, and wildlife. But in due course I asked questions about forestry and land use they could not answer. Forest Service photograph.

In August 1966 the Forest Service invited me to join a field study trip into the High Uintas Primitive Area in central Utah in preparation for proposing wilderness designation for this unusual east-west mountain range. And here we are after a week in the high country. *Left to right:* unnamed Forest Service officer; Stewart Brandborg, executive director of the Wilderness Society; Michael Frome; John Mattoon, Forest Service information officer; John Herbert, strong wilderness advocate in the Forest Service; Clifton Merritt of the Wilderness Society; and Carl Hayden, reporter for the *Salt Lake Tribune.* Forest Service photograph.

At the foot of the glaciers in 1965, Superintendent Bob Howe was glad to share the wonders of Glacier Bay National Monument, Alaska (later reclassified as a national park). Neither he nor I had any idea that in a few years seemingly remote Glacier Bay would be subjected to intrusion by many huge cruise ships. Photograph by Bob Spring.

In early spring 1983, while hiking the Alum Bluff Trail to Mount LeConte in the Great Smoky Mountains, we were caught in an icy early spring blast, but we pushed on. Photograph by Mack Prichard.

On the trail in the Washakie Wilderness, Wyoming, in 1993, I realized I was older than I used to be, so Carissa, my friend's dog, and I studied the high country from a resting position. Photograph by Lynn Kinter.

Robert Cahn and I covered much of the same turf as journalists, but we found plenty of room for us both and cheered each other. Cahn won a Pulitzer Prize in 1968 for a penetrating series in the *Christian Science Monitor* on the problems of national parks and in 1970 was appointed one of the first members of the President's Council on Environmental Quality. He is shown here in 1981 autographing his book *Footprints on the Planet*. Photograph by Mack Prichard.

Another Pulitzer Prize winner, Tony Auth, illustrated my column in every issue of *Defenders of Wildlife Magazine* for more than fifteen years and sent me a gift of this original cartoon from the February 1983 issue. (The column was titled "Many Idahoans Give Wildlife a Chance Despite Their Political Conservatism.")

Brock Evans, while vice president of the National Audubon Society, is shown in a typical upbeat pose at the annual Audubon conference at Estes Park, Colorado, August 1991. Brock went west from Ohio early in the 1960s to practice law in Seattle. Love of the outdoors, however, led to appointment as Northwest representative of the Sierra Club, later to Washington, D.C., and Audubon, then as head of the Endangered Species Coalition. Photograph by Mack Prichard.

Tom Bell, founder of *High Country News,* gave heart and voice to the environmental movement of his home state of Wyoming and the West. We are pictured here at Whiskey Basin, Wyoming, with his wife, Tommy, in 1993. Photograph by Lynn Kinter.

We already were friends and worked together for years when Robert Stanton, a regional director of the National Park Service, came to lecture at Western Washington University in 1992. Later he was appointed the first black director of his agency and served until the election of George W. Bush. Photograph by Mike Wewer.

Rupert Cutler, after a long private and public career, came for a visit in 2000. He noted that we both were older and heavier, but we still had such a good time that I picked up the check.

Mack Prichard, the intrepid Tennessee naturalist, and I have laughed and cried and never looked back while traveling many environmental trails together. In this picture we were in Cincinnati in 1993 when I received my doctoral degree.

"The Wild Bunch" was the headline over this picture when it appeared in the *Missoulian* in Missoula, Montana, early in September 1994, when we gathered at the Craighead Wildlife-Wildlands Institute to celebrate the thirtieth anniversary of the Wilderness Act and, as the newspaper noted, "to ask for a reaffirmation of the wilderness ideal." *Seated, left to right:* Paul Fritz, Michael McCloskey, David Brower, Ann Brower. *Standing:* John Craighead, Stewart Brandborg, Michael Frome. Photograph by Michael Gallacher, the *Missoulian.*

Hugh Iltis, professor of botany emeritus at the University of Wisconsin, shown here in his front yard at Madison in summer 2001, has been a friend and collaborator of sorts for thirty-plus years. Iltis, addicted botanist and "half-plant," has exhorted fellow scientists to pursue activism as he does. Photograph by J. Eastvold.

At the Sigurd Olson Institute for Environmental Studies at Ashland, Wisconsin, in summer 2001, I paid tribute to a friend and exemplar. Olson was a highly effective speaker (as well as writer), reflecting his training as teacher and guide. At public hearings in northern Minnesota he withstood the wrath of neighbors eager to commercialize the canoe country. Photograph by J. Eastvold.

In the Northwest, late 1990s.

Speaking, Teaching, and Raising a Little Hell in Academia

In 1978 I was invited to spend half a year as a visiting professor of environmental studies at the University of Vermont. I had lectured and given seminars at colleges and universities here and there, but being a regular in the classroom was a new experience.

Vermont marked the start of still another career. I hardly anticipated the challenge of rowing upstream, or perhaps I should say of skiing uphill, in higher education, and of pulling up my roots, such as they were, to become a bit of a rootless itinerant. Luckily, I found adventure and reward in every place I've lived.

I loved Vermont, even though winter was an endurance contest. Two feet of snow fell on two feet of snow, but life went on. Observing Vermonters persevere in their harsh environment was a lesson in itself. I took up cross-country skiing. I fell down but learned to use skis and poles to pick myself up and move on. Ski areas were all over the state, and every weekend I tried a different one. I've long forgotten many of them but treasure the memory of skiing on frozen Lake Champlain in quiet moonlight. In the end, at Blueberry Hill, I bought a used pair of Norwegian-made wooden skis and the appropriate waxes, all destined to become quaint relics, but I used them for years whenever I went where there was snow.

I found more villages than cities, each looking like a picture out of a book of old New England, but clearly unsuited for making much of a living. Many natives were hardy, patient, and maybe insular by choice. City people who had moved there were about the same. One fellow I met, a reformed New York lawyer, preferred to dress down rather than dress up and to ski out of his front door instead of running for the subway.

You don't go to Vermont looking for issues, I reminded myself more than once, you go there to get away from issues. I mean that at first I was shaken by what I found and felt at the university. It seemed to me that little rich white ladies and gentlemen were getting a C+ education to go out and become C+ consumers. Then I reasoned that after the turbulent Vietnam protests of the 1960s and Earth Day 1970, students simply had calmed down and were no longer aggressive.

Maybe it was the influence of their professors. I made many friends in academia over the years, but there's something haywire about the system. At Vermont, professors kept their doors closed. They weren't social or helpful. Although professors refer to their "collegiality," they get nasty about each other, with smart cracks like "He writes books for children that demean the status of professor. They're not for adults." I met a professor who had just been denied tenure and was absolutely shattered, especially on reading in his evaluation: "He attracted intellectual students. Students gave him mixed response to an uneven performance as a teacher." A new young professor assigned to teach a "gut course" (where students are expected to report in body only, leaving minds at home) told me of overhearing one student complain to another, "This guy expects us to read a book!"

Environmental studies were mostly about theory and philosophy. Students were searching for idealized concepts of nature rather than how to face reality. Soon after my arrival I came to lecture in a large class. The students were ready to tell me that activism was about as repugnant to them as communism. One demanded, "Why have confrontation?" When I suggested that we need to eliminate affluence and waste, one piped up, "Are you promoting poverty?" And another asked if I was motivated by a guilt complex.

I'm sure that at Vermont and other institutions to follow I received "mixed response to an uneven performance." My courses were not well structured, like those of colleagues, who relied on time-tested outlines and notes. I wanted to bring my own experiences in writing or in involvement in issues, or the day's headlines, or something spiritual, or something absurdly unrelated, into the classroom, and to take students out of the classroom into life. For many students, it was not what they expected or wanted.

On the other hand, I found a cadre of students ready and willing to be involved and to learn by doing. In one class, Eric Swanson was in a class by himself, involved in campus issues and local issues, and in behalf of the Alaska Coalition for the protection of public lands under question, and still doing excellent work in his reports. "If you don't watch out, Eric," I told him one day, "you'll end up with an A-plus." "I'm not here for grades," he snapped back. "I'm here to learn to be effective."

In my other class, I had Betsy, Cynthia, Dana, Rick, and other bright young people aspiring to serve the public good. At that very time some rural Vermonters and politicians were promoting a bounty on coyotes and hunting of bobcats in order to reduce supposed predation on domestic sheep. It was the old unscientific and wasteful approach more likely found in sheep-growing states of the West. We discussed it in class and decided to conduct an evening program, inviting as guest speakers Toby Cooper of Washington, D.C., representing Defenders of Wildlife, and Ted Williams of neighboring Massachusetts, a well-known outdoor writer who formerly worked for the Massachusetts state wildlife agency and would later be admired for his work for *Audubon* and other magazines. Students wrote their own publicity materials, contacted the media, invited state officials from Montpelier, and ran a lively evening seminar that attracted more than two hundred people.

I thought it was a good show, but presently David Capen, director of wildlife biology in the university's school of natural resources, dispatched a hot memorandum to my department head, declaring:

> Ted Williams blatantly attacked state fish and game agencies, both individually and collectively, and degraded wildlife biology as a profession. He presented a distinction between a "wildlifer" and an "environmentalist" and continued to expand on that cleavage. I, for one, have naively subscribed to the notion that a wildlife biologist is an environmentalist.
>
> Our program endeavors to train students so that they are qualified to obtain employment with fish and wildlife agencies, and we attempt to cooperate in many ways with the Vermont Fish and Game Department. Certainly you can see that any association between our program and the presentation entitled "Defenders"

could endanger relationships with the agencies which
we attempt to serve and complement.

Of course, Professor Capen and colleagues were training students to
work in state game agencies, where the word "defenders" was anathema.
And years later the chair of the journalism department at Western Washington University would sharply remind me that students were being
trained to work for local daily newspapers, where "objectivity" in reporting presumably prevailed and that I should not suggest otherwise.
Maybe not, but I did, and I do, believing a teacher should broaden, not
narrow, the student's field of vision.

At Vermont, Rick Simonsen, an excellent and promising student,
corresponded with Ted Williams, who wrote him:

> It is not I—or anyone else—who has degraded
> professional wildlife biologists. They've degraded them-
> selves. You and I both know that biologists do not
> want to continue ridiculous programs such as stocking
> cock pheasants. The good ones, that is. But the point
> is that they don't have the guts to fight these pro-
> grams. It's much easier to sit back and blame the politi-
> cians, the powerful sportsmens groups, and tradition.
> After six years of working with biologists, I don't feel
> awfully sorry for them.

In another project in the same class, students dared to challenge the
local public utility, Burlington Electric, as it prepared to implement an
absolutely frightening plan to fuel its plants with wood chips logged
from Vermont forests. At first, nobody paid attention to the students, but
they conducted serious research and got the word out to the community. Finally, the *Burlington Times-Argus* reported on May 4, 1978:

UVM Activists Protest Planned Power Plants

> A group of environmental students at the University of
> Vermont Wednesday issued the first sharp, in-depth
> criticism of the proposed series of new power plants
> planned by the Burlington Electric Department.
>
> The plants in question have to date won nothing
> but praise from virtually everyone familiar with them,
> because they appear to make innovative use of three

energy sources not tapped on a wide scale in Vermont: wood power, trash power, and water power.

But the student group issued a series of fact sheets Wednesday which were sharply critical of the proposed facilities.

The student group said the proposed new plants would probably cost significantly more than estimated, might not work properly, and could cause significant environmental damage.

They urged a moratorium until all questions surrounding their construction can be answered.

Ultimately the city prevailed, but the students did their homework; they made people think and question authority, and they themselves felt they had learned by participating in the process of decision making. (Later, the Sierra Club caught on. It persuaded a judge to limit the amount of pollution at the plant. The restrictions applied only to wood burning, and the plant switched to polluting oil and natural gas.)

When my teaching was done in Vermont I returned home (to a Virginia suburb of Washington, D.C.) and forgot about teaching. Then in 1981, with a marriage failing at hand, I faced a variety of problems and wanted to get away again. I was trying to prepare a new edition of *The Forest Service* (published first in 1971), a book I had written as part of a series on government departments and agencies, but without making headway. One day, while working through my dilemma, I had lunch with the chief of the Forest Service, Max Peterson, whom I had known through much of his career. We may have disagreed on issues, but Peterson never tried to silence or avoid me. While we were eating and talking, he said it wasn't easy to run an old-line agency or "turn the ship around." For example, he cited a letter from a prominent retiree: "Dear Max: With reference to that speech you gave on multiple use and wildlife, why don't you cut out that bullshit?" Then the chief surprised me by suggesting that I go to the Pinchot Institute to work on my book.

Thus he made it possible for me to spend a year at the Pinchot Institute for Conservation Studies, at Milford, Pennsylvania, the converted forty-room mansion Gifford Pinchot had inherited from his wealthy parents and then bequeathed to the Forest Service. There really wasn't much in the way of other conservation studies, but I had comfortable working space and spent free time exploring the bordering Pocono

Mountains and Delaware Valley and learning more about Gifford Pinchot. He was a patrician motivated by *noblesse oblige,* a close ally of Theodore Roosevelt early in the twentieth century, first chief of the Forest Service until he was sacked by Roosevelt's successor, William Howard Taft, and then twice governor of Pennsylvania, He was a Republican social crusader of another age, full of ideas about progressive forestry and conservation until he died in 1946 at the age of eighty-one.

The Pinchot mansion, Grey Towers, built in the style of a French country chateau, was bordered by a huge patio ideal for summer dining. One evening Max Peterson came up from Washington to speak to a dinner meeting of a Pennsylvania chapter of the Society of American Foresters. The booze flowed and by the time the program began the chapter chairman, an official of a large timber company, was half-loaded or worse. In his introduction of the chief, he rambled on for thirty minutes or more with a stream of crude, vulgar, dirty jokes. I watched Peterson. He didn't laugh, but he didn't grimace or show displeasure or impatience, as I might have. He waited with seeming patience and delivered his remarks with good grace. It was not a pleasant event, but there was a lesson there that might prove difficult to share in the classroom.

Another day that summer Dale Jones, then director of wildlife and fisheries of the Forest Service, telephoned from his office in Washington, D.C. He understood that I was working on my book and wanted to visit and discuss it. When he came to Grey Towers Jones expressed his concern about protecting wildlife on the national forests. He spoke of the need within his agency of "ecological perceptions as the basis of responding to a demand for the peace of mind that comes from protecting and preserving wildlife because its users see it as a treasured gift." I certainly did try to incorporate that idea into the book, and into teaching, writing, and speaking. Ecological perceptions and preserving wildlife as a treasured gift ought to be paramount in the goals of professionals working on public lands and of a lot of other people as well.

When my year at Milford neared its end, I cast about again. I did not expect another chance in academia; I was hoping for someplace else like the Pinchot Institute, but another Forest Service friend, John Marker, on my behalf contacted James R. Fazio, in the forestry school at the University of Idaho, who knew of my work and was very receptive. As chairman of the department of wildland recreation, he couldn't offer a lot of money, but he offered what he could and in due course became a very close friend.

I felt blessed anew, having been in Idaho before. I had come in the 1960s to observe and write about the political battle to save Hells Canyon of the Snake River, a marvel of the planet, deeper than the Grand Canyon of the Colorado River. I came again in the 1970s to the dedication of yet another marvel, the Birds of Prey Natural Area. I knew the battle over the River of No Return Wilderness, when Senator Frank Church conducted hearings in different sections of the state, inducing people who had never spoken publicly before to stand and open their hearts in praise of an area larger and wilder than Yellowstone. The River of No Return, covering 2.2 million acres, was not only the largest unit of the National Wilderness Preservation System outside Alaska but also the core of a great network of Idaho wilderness, including the Gospel Hump and Sawtooths and the Selway-Bitterroot extending into Montana.

When I arrived at Moscow, Idaho, in autumn 1982, I found the university maintained on its campus a Wilderness Research Center, whose functions included sponsoring an annual Wilderness Resource Distinguished Lecture. Before the year ended I was honored with the invitation to deliver the sixth annual lecture in the series.

I elected in my speech to talk about patriotism, citing attacks by Reaganauts running the government on the patriotism of environmentalists. Interior Secretary James G. Watt called environmentalists "a left-wing cult which seeks to bring down the type of government I believe in." Assistant Secretary of Agriculture John B. Crowell (a timber industry attorney placed in charge of national forests) said he was sure the Sierra Club and National Audubon Society were "infiltrated by people who have very strong ideas about socialism and even communism." A report by the Republican Study Committee in the House of Representatives, warning against the "specter of environmentalism" and the "hidden liberal agenda," declared, "Environmentalists are liberals and self-motivated and intent on preserving their privileged social status."

I feel it is morally wrong when officials in high places, or in any places for that matter, try to limit debate by putting labels on those who disagree with them, who deride and impugn the motives of those who want to be heard. I cited the case of my friend and hero, John P. Saylor, who introduced the Wilderness Act in the House of Representatives and fought for it until it passed. He was a Pennsylvania Republican, certainly not "liberal" or "progressive," but motivated by good, old-fashioned patriotism. He was never divisive or destructive, never challenged the motive nor questioned the loyalty of those who disagreed with him.

I was well received. I went to potlucks with faculty colleagues and their spouses; hiked, backpacked, and skied with colleagues, students, and a girlfriend; bought a used bike and rode around town; and kept writing in my little apartment. One April, students and I celebrated Earth Day on campus with music of dulcimer, fiddle, and guitar and discussion of local issues.

I made a few speeches at home and away. In April 1984 I talked on campus at Natural Resources Week, a kind of homecoming celebration held by the various forestry schools when they would welcome back their alumni to show undergraduates the fruits of their future. It was also a time for timber companies to renew gifts of money and equipment and to remind the schools of how much they had given and how much they are owed. Two weeks later the dean and my department head received an aggrieved letter from Bruce E. Colwell, an official of Diamond International, a timber company of Coeur d'Alene, Idaho:

> To say that I was surprised at Mr. Frome's remarks is putting it mildly. I was totally shocked that such a talk would be made to a group of future resource managers. It is my firm belief that it is being totally irresponsible to suggest to a group of professionals that they become activists as was suggested in the talk by Michael Frome in order to counter the activities of the large companies. It is inferred by you that to offset this it is advisable to have someone to present the other side of the issue and let the students determine what is right.

The problem was that to most of the timber industry there was no "other side" to counter what the industry liked to call "multiple-use resource management." It was astonishing in a way that I should be welcomed to join a forestry school and very gratifying that the administration resisted industry efforts to get rid of me.

I stayed at the University of Idaho four years. I wasn't obliged to leave, but felt it was time. I was not on a tenure track and really had no professorial status. The dean wanted me to stay but had little to offer. Fazio had moved "upstairs" to be associate dean, and the focus of the department had changed from wildland preservation and recreation to recreation and tourism. My colleagues and students tendered a farewell party with emotion and good feeling, presented me with a handsome Pendleton shirt that I still wear occasionally and announced establishment of the Michael Frome Scholarship for Excellence in Conservation Writing, a prize awarded annually to a deserving undergraduate or graduate student.

In 1986 I moved east, or at least eastward, for a one-year connection at Northland College, at Ashland in northern Wisconsin, where I was impressively titled environmental scholar-in-residence and provided with a pleasant office in the Sigurd Olson Environmental Institute, a kind of outreach division of the school.

I had spoken at Northland a few months earlier at a conference on wilderness. The administrators evidently liked my speech enough to talk with me about joining the staff for a while, and forthwith I agreed. I had known Sigurd Olson before his death in 1981 and felt it a privilege to associate with an institution named for him. Olson was an environmental hero in the upper Midwest, who lived for years at Ely, Minnesota, at the edge of the Boundary Waters Canoe Wilderness, but he had lived as a boy in Ashland, where his father was a minister. I had known him as both an author of noteworthy books about his adventures in wilderness and as an eloquent activist and vice president of the Wilderness Society.

The college at this time was down in enrollment, with less than six hundred students. Advertising and catalogue described the college as "a liberal arts/environmental college," but outside of the Olson Institute I did not see much at all to the environmental part of it. Most of the faculty seemed indifferent; there was nothing environmental integrated in the curriculum, and very few of the faculty joined in activities of the institute.

The institute, however, did good work, conducting educational programs in nearby northern Minnesota, Wisconsin and the Upper Peninsula of Michigan, distributing constructive materials about wolves, loons, and caring for the earth. I was pleased to associate with bright young colleagues, who cared about me too. During the summer, at the outset of my stay, Mark Peterson, the institute director, and I drove all the way around Lake Superior, the largest and cleanest of the Great Lakes, exploring beautiful landscapes and vistas, like the multicolored sandstone cliffs called Pictured Rocks rising steeply along the Michigan shore of the lake and Pukaskwa (pronounced Puck-a-sah) National Park, on the Ontario eastern shore, preserving a fragment of Canadian Shield forest wilderness and ancient Indian rock carving. On my own, or with colleagues or students, in due course I visited the Apostle Islands, close at hand off the Bayfield Peninsula, and Isle Royale, the roadless national park in the heart of the lake.

As part of my work for the institute, I conducted what we called the Sigurd Olson Memorial Wilderness Lectures, a well attended series featuring national environmental leaders who came virtually without expense or fees—although we did pay something to Dave Foreman, the flamboyant Earth Firster, who needed it more than the others.

Even before Foreman's arrival, a clutch of students had organized what they considered an Earth First chapter. They debated whether to be a legitimate campus club or go underground and undertake "monkeywrenching" in the dark of night. They kept coming to my office for advice, but I told them I didn't want to know anything about that part of it. Then the college delivered an issue that enabled the students to do something daring without doing damage.

Someone had discovered maintenance personnel dumping debris into a choice wooded natural area on campus called the Ravine. One night the Earth Firsters hauled out a collection of mangy rubber tires, broken old machine parts, and assorted other garbage and built a display of it all, mounted with blazing banners, in the middle of the campus. In the morning the president of the college, Malcolm McLean, arrived on the scene, thunderstruck and embarrassed. He exclaimed, "If they had only come to me! I would have come and worked with them to clean it up. Of course I will ensure that the debris is disposed of properly."

McLean was mature, tall and fit, a conscientious, progressive administrator who seemed sincerely concerned about students as individuals and to know many by their first names. He was fair and friendly to me and I enjoyed discussions with him. After a particular speech I delivered, he wrote to me:

> Your call to idealism, to have people rise above the norm, to emulate the example of Sigurd Olson and to make our own distinctive mark in the world, is in my judgment just right. It's a message that needs to be repeated again and again.
>
> One small point. You referred to a "college administrator" who used statistics showing a declining interest in environmental concern. My point in bringing this up at your earlier presentation here at the College was in no sense to be discouraging. It is simply that we need to understand the ambiance in which we are operating. The young people who are the hope of the

world are idealists. There is not a great tide of them as there seemed to be during the Earth Day years. Now, increasingly they are a minority and I think those figures show that they are. To disregard them altogether seems to me to misread a bit of the world in which they must operate. In no sense was citation of those statistics a condemnation of the young, nor a call for defeatism. Far from it. Rather they must understand that they are to some degree going to be part of an embattled minority—just as Sigurd Olson was, and just as John Muir was. That's all.

I responded that, yes, we do need to understand the ambiance of things:

But what do figures mean? Figures tell me that millions of Americans watch sports, sex, and crime on television. Figures tell me that young people take drugs and cannot spell. But the world is the way I want it to be, and I don't want it to be mediocre.

Attached is copy of a questionnaire to which Lynn Kinter, age twenty-four, one of our wilderness series lecturers [a former student of mine at Idaho, working to preserve western wild desert], responded. Note her assessment of the change in young people and her further deduction: "I think it is not only young people's attitudes that have changed, though. At times it seems as if our nation has become one largely populated by zombies."

Surveys and statistics may even show the same, and they may be important in merchandising. But my nation is not populated with zombies, whether with or without college degrees. Lynn predicts a resurgence of public interest in conservation. Surveys may indicate it, and institutions appropriately respond and once again offer environmental activities and studies. I'm not waiting, nor worrying about being "an embattled minority."

In retrospect, if my view was valid, so too was his. He had to run this little college while I had the luxury of being able to criticize. He gave me the platform from which to speak my piece.

From Northland I moved to Western Washington University, where I established a course of study in environmental journalism. I stayed for eight years until 1995. It was the longest period in any job since the American Automobile Association. I was seventy-five, still feeling energetic and productive, but it was time to get out and get on. I left to retire and to enjoy my marriage (to June Eastvold on New Year's Eve 1994). My wife, an ordained, practicing Lutheran pastor in Seattle, was wholly committed to her faith and yet with a holistic religious feeling matching my own. She retired from her position soon after I left mine. In 2001 we still lived in the house I bought when I came to Bellingham, Washington, in 1987. It wasn't that I was through traveling, but now I had a companion and a home to come home to.

About speaking in academia. I hope I made the most of my opportunities. The best of it was working with students, helping them on their way. I lost track of most, but not all. Eric Swanson, at the University of Vermont, whom I mentioned earlier, came to Washington, D.C., to work for the Izaak Walton League, then in other positions, and in 2000 became vice president of Common Cause. Scrawny little Jeanette Russell, at Western Washington, could not have weighed more than 110 pounds dripping wet. Her writing was weak and her spelling was awful, but she was a determined activist who after graduation went to Missoula, Montana, to make her way. While there for a conference, I had lunch one day with Jeanette, who acted with poise and confidence and was married to a very nice young fellow. She showed me the printed newsletter of the National Forest Protection Alliance, which she edited. I couldn't believe it, but it looked great. Jeanette had actually helped organize this alliance of grass-roots groups all around the country and was its executive director. "Jeanette," I asked, "what grade did I give you?" To which she replied, "You gave me the only A I ever got." She earned it, and we could both feel proud.

A Curriculum for
Wilderness Leadership

At the Sigurd Olson Environmental Institute,
Northland College, Ashland, Wisconsin,
April 7, 1987.

In September 1994, Stewart Brandborg, John Craighead, and I initiated a celebration of the thirtieth anniversary of the passage of the Wilderness Act, which we held on the grounds of the Craighead Institute for Wildlife Research in Missoula, Montana.

For the morning program a panel of old-timers reminisced about personalities and events of the early sixties leading to passage of the act, and for the afternoon a second panel of young activists discussed current issues and controversies. Next morning's edition of the *Missoulian,* the local daily, carried a large photo of a group of us beneath a bold headline, "The Wild Bunch," and accompanied by a comprehensive article written by Sherry Devlin.

In the body of the article Devlin reported an interview she conducted during a break in the proceedings with David Brower, John Craighead, and me. Brower was gloomy. He saw the world in a downhill spiral, without much hope. It was understandable—I remember him saying to me, either at Missoula or somewhere else, "Everything is out of control"—but out of sync with his life's work. Craighead and I were more optimistic, insisting the public ultimately would awaken and turn things around.

It wasn't exactly easy for Craighead to be optimistic. He was a wildlife scientist who knew more about grizzly bears than anybody and was obliged to watch painfully over the years as their roaming room and their numbers declined. With his twin brother Frank, John Craighead conducted pioneering grizzly bear research in Yellowstone National Park, utilizing telemetry to track bears and learn what they do and where and when. The Craigheads were nationally known, featured in *National Geographic Magazine* and in National Geographic films. But the National Park Service grew unhappy with them, perhaps because they were *too* prominent. Jack Anderson, the superintendent of Yellowstone National Park, said he did not want park visitors looking at bears wearing radio collars, which he considered unnatural in a natural setting. The Craigheads' research permit was canceled and they and their research team were dismissed from the park.

Paradoxically, Superintendent Anderson received the very first award of merit from the International Snowmobile Manufacturers Association for his personal leadership in introducing recreational snowmobile use to Yellowstone and in welcoming snowmobilers from all over the country. He said it would prove to be a very good thing and that snowmobiles would not in any way bother bison. Time has proven snowmobiling to be a very bad thing for Yellowstone and its bison, so very bad the National Park Service in the year 2000 moved to eliminate it completely.

It wasn't easy for Brandborg to be optimistic either. Elsewhere in these pages I refer to his career in the Wilderness Society. Following his dismissal as the society's executive director, he joined Jimmy Carter's presidential campaign staff. Working from Atlanta headquarters, Brandborg organized Conservationists for Carter, tapping into the network of grass-roots activists he had earlier established to work for wilderness. Carter won a slim victory over Gerald Ford, and for all we know the conservationists made the difference.

After the election, Brandborg joined the transition team at the Department of the Interior, selecting, or at least recommending, appointees for key positions. He expected that in due course he would be one of them. I recall that we talked frequently on the phone during this period and that I shared his confidence. So I was privy to the sequence, as follows:

Cecil Andrus, the governor of Idaho, telephoned Brandborg and met with him in Washington. Andrus asked if Brandborg was himself campaigning to be secretary of the interior and, if not, would he support Andrus. In due course Brandborg agreed, led to believe Andrus would ensure a key spot for him. Then appointments were announced, first at the higher levels, the undersecretary, solicitor, and assistant secretaries, then at the second levels, the bureau chiefs. Brandborg was passed over for them all. He tried to see Andrus but couldn't get near him. He went to see Representative Morris Udall, chairman of the House Interior

Committee, with whom he had worked closely while at the Wilderness Society. Udall interceded in his behalf, and Brandborg was appointed an assistant to the director of the National Park Service, where he did good work trying to train personnel to work together in trust and to communicate openly with citizens who care about national parks.

In the speech below I talk about a curriculum for wilderness leadership. Personally, I learned a lot hanging out with people like Brandborg and Craighead—not just hanging out, maybe that's not the right way to put it, but working with them, as I was privileged to do. So, for vision and mission, I would definitely include that kind of experience in the curriculum.

■   ■   ■   ■

It is important always to hold aloft the *why* of wilderness. In a few words, wilderness embodies freedom, challenge, and inspiration. Wilderness is a living document of land and people, as valid and vital as the Constitution, the two-hundredth anniversary of which we celebrate this year. Wilderness preserved marks humankind's respect for the earth and for itself.

That much is easily stated. But *how* can wilderness be saved in our overpopulated, supercivilized age, dominated as it is by technology and television? Where can we possibly find leadership for nature protection and preservation, when able young men and women are drawn to career futures in computers and commerce? Even at best, where can those who care find the training they need? Where is a curriculum for wilderness leadership?

Over the years I have explored the wild places, read, written, and lectured about them. I must say, however, that in all my adventures I have never come across an institution of higher education equipped to provide leadership in wilderness. That simply isn't what education is about. Education, with only a few exceptions, is about careers, jobs, success in a materialistic world, elitism, rather than caring and sharing; it's about facts and figures, cognitive values, rather than feeling and art derived from the heart and soul; it's about conformity, being safe in a structured society, rather than individualism, the ability to question society and to constructively influence change in direction.

A change in direction is critical and imperative. The most important legacy our generation can leave is not a world at war, nor a nation in debt to support a nuclear Star Wars system, nor the settlement of outer space, transporting all our worldly problems to the rest of God's universe, nor the breeding of test-tube babies and robotic drones. Our most precious gift to the future, if you will ask me, is a point of view embodied in the protection of wild places that no longer can protect themselves. Conservation education thus must enlist not only rational recognition of the problem, but human concern, distress, and love. We must understand more about the history of ideas that dominate the philosophy and policy of society, that dictate our obsession with facts and figures, more about the analytical type of thinking of Western science that gave us such power over nature while smothering us in ignorance about ourselves as part of it.

"But let children walk with Nature," wrote John Muir, "let them see the beautiful blendings and communions of death and life, their joyous inseparable unity, as taught in woods and meadows, plains and mountains and streams of our blessed star, and they will learn that death is stingless indeed, and as beautiful as life, and that the grave has no victory, for it never fights. All is divine harmony."

Muir wrote those lines after sleeping in a cemetery in Georgia during the course of his legendary thousand-mile walk to the sea. We all ought to sleep in the cemetery and listen to the voices there. I feel that I should, and that I should teach in the out-of-doors classroom, absorbing divine harmony as the basis for restoring harmony to human society.

While in California two years ago, I revisited condor country, the last slender sanctuary of hope for an American species at the brink of extinction. I had been there before, in the foothills and mountains north of Los Angeles, and have felt that saving the condor is like saving the whale or the redwood tree. The very process of doing so enriches civilization; it fulfills the human obligation to other species that share the planet with us. Furthermore, if we can save whales and redwoods, then we can save condors as well.

"The California condor," wrote the late Carl Koford, "is a majestic bird in its natural rugged environment as it sweeps in superbly controlled flight over crests of ridges and great slopes of tangled chaparral. The air passing through its wing tips sets up a steady whine as it is pressed into service to keep the great glider aloft. The condor passes overhead, the sound recedes, and the bird now circles and scans with keen eyes the ground below and the activities of its fellow condors."

I cherish this word picture which I first read more than thirty years ago. It enables me to visualize a bird unmatched by any on this continent in its wingspread. Koford's words also enable me to appreciate the value of research and of committed researchers like himself. During his field studies from the 1930s to the 1950s he watched the condor day and night with endurance and patience; he went where there were no trails—backpacking, cave-dwelling, climbing, and observing. When he departed the Topatopa Mountains, he carried with him the story of a vanishing American, struggling for survival in its last shrunken stronghold. But Carl Koford realized that his findings were only part of a pattern, that research, or "science," hardly stands alone. I have always been wary of those who say, "Trust me, believe me, let me chart the course because I am equipped with scientific training." The challenge of the condor, as the challenge of wilderness, embraces philosophy, ethics, political science, and communication, as well as technical research. Above all, it takes a lot of caring about condor country and commitment to its preservation and appropriate use, or no use at all, by humankind. This goes beyond the curriculum of science.

I pursued this idea when I went on from California to Hawaii. While climbing to the summit of Mauna Loa, the world's largest volcano, a mountain built by layer upon layer of lava, I thought of the early Hawaiians making their way to the top without shoes, backpacks, or freeze-dried food, perhaps without warm clothing, living close to nature and free of the artifices that clutter our advanced civilization.

I hiked upward through a forest of ohia, the pioneer tree of fresh lava flows, and heard the rambling, rolling song of the apapane, the most common surviving species of Hawaiian honeycreeper, flitting from one tree to another in the forest canopy to feed on nectar from ohia blossoms. Everything natural and native about Hawaii—its birds, insects, plants, ferns, and trees—seems so distinctive and luxuriant.

And yet all of it is acutely vulnerable. Of the seventy bird species found nowhere else in the world, twenty-nine are on the threatened or endangered species list and at least three are close to extinction. Why should this be? Andrew J. Berger, the retired chairman of the University of Hawaii's Zoology Department, is the foremost authority on Hawaiian birds, with many field trips over the years across the Hawaiian chain. Dr. Berger brings credit to ornithology by refusing to hide behind scientific jargon, instead placing blame where it belongs in language anyone can understand.

As he relates in his definitive book *Hawaiian Birdlife,* in the 1950s the Hawaiian Division of Forestry and U.S. Forest Service promoted destruction of native koa forests in order to plant pines. "Until recently," he writes, "the board-feet-oriented federal foresters referred to the endemic Hawaiian ecosystem as 'decadent forests,' consisting of 'weed species' and 'unproductive forest land.' Consequently, state forestry plans have continued to place emphasis on planting exotic trees."

Dr. Berger's assessment is bitter and unequivocal: "That the Hawaiian biota should have been raped, ravaged and devastated during the 19th century was regrettable even though understandable, but that the rape has continued not only into the twentieth century but even into the eighth decade of that century is a sad commentary on man as an animal species. Man is, indeed, a disease on the planet earth."

I hope not, though the visible record of our time, in Hawaii as elsewhere in the world, is clear enough to support the charge. Our friends, the foresters, like other technicians, are victims of their training, a pseudo-science focused on commodity production and commercial values, treating the earth like a mere resource to be exploited rather than like a mother to be honored and cherished. But we are all victims of the same system.

Something more fundamental is needed than the education of today. "Higher education in America is suffering from a loss of overall direction, a nagging feeling that it is no longer at the vital center of the nation's work. After decades of enthusiastic growth, many colleges now face confusion over goals, reduced support, and an uncertain future." These are not my words, but those of Ernest Boyer and Fred Hechinger, from a report published by the Carnegie Foundation for the Advancement of Teaching.

I myself would go further. Native Hawaiians speak of "aina," the traditional love of land. Their poetic oli, or chants, and the hula recount stories and traditions of humankind woven into the natural universe. The summit of Kilauea is considered sacred, the palace of the goddess Pele. As daughter of Earth Mother and Sky Father, Pele came to Hawaii in flight from her cruel older sister, the goddess of the sea. She found her refuge at last in the volcano, where she has prevailed ever since as goddess of fire. This world view may be dismissed as superstition or respected as reverence for life. Take your choice, but while silence may increase knowledge, scientific data cannot be equated with feeling that derives from the heart and soul.

Both science and spirituality are needed to save our wilderness and ourselves. At a 1984 symposium on Hawaiian ecosystems, a report by two scientists, Alan Holt and Barrie Fox, acknowledged as follows: "If a thousand years from today there are large areas of native landscape in

Hawaii, it will be because people cared enough to save them, cared enough to keep natural areas protected even in the face of other potential uses of these lands. The long-term success that we all hope for depends on the people's appreciation of the land. The best prospect for making that future happen is to show today's people the value of our natural heritage and to show them how to care for it."

But who will do the showing? Where will those who truly want to save wilderness find their curriculum for leadership? I feel that we need a revolution of ideals, a revolution of ideas in all fields. We need a revolution of ethics to sweep America and the world, because the same forces are at work everywhere. We must alter the superconsumptive lifestyle that makes us enemies of ourselves, a life-style that confuses a standard of living with a quality of life.

That kind of revolution begins with the individual, inside oneself, with one's own ecosystem, finding the unity of body, mind, and spirit and reaching out to others to do the same. That is where constituency building best begins. But there must be a lot more to it than wholesome personal living. The curriculum includes learning to write, speak, research, analyze, participate in the process of decision making, reach independent judgments, take risks, listen to everyday people and to understand their needs and goals—all of these are needed in the training of wilderness leaders.

My daughter Michele was involved in environmental studies at college in New England. After graduation she worked for a time for the Vermont Natural Resources Council, then enrolled for a graduate degree in government at Harvard. One day she spoke in her class on the principle and purpose of Vermont town meetings. When she was through a classmate poked fun: "How can you speak of town meetings? You're at Harvard!"

Actually it could be any institution training professionals, as well as Harvard. With a cadre of experts, after all, who needs the people in decision making? The truth is that most experts play it safe. That is why they are considered experts. They don't challenge the system when they are part of it.

A couple of years ago I was invited to discuss conservation before the Seattle Garden Club, a large group with socially prominent members. Before the meeting I was informed that one member had inquired, "Is he really going to tell it?" while another wondered, "Will he be too political?" The conservation chairperson asked me, "Could you provide a serious message, but without offending anyone?" And I received a letter from yet another member, warning that "with ladies, conferences and studies are just great, but actions scare them silly, especially if their husbands and friends don't like it." With higher education, conferences and studies are great, too—but actions scare them silly, especially if their peers and sources of funding don't like it. To be involved is not "professional." You lose "credibility." Few are big enough to break out—few in any field, whether in business, Congress, politics, the media, the professions, or education.

I enjoyed four glorious years at the University of Idaho. However, I recall this admonition from my department head: "But, Michael, you're not training students to think; you're teaching them to be like you. . . . You're opinionated. You should provide the data and allow students to reach their own conclusions." Such is the classic academic approach, whereby the instructor is presumed to have no opinion and thus issues no challenge. I recall a particular incident when a student came to me to turn in a paper. He was puzzled. "I don't understand the assignment," he said. "It made me do a lot of thinking, but I don't get it. Is there a right answer?" Undergraduates are conditioned to multiple-choice or true-false exams, memorization and regurgitation, where there must be a "right" answer, not his or her own answer. Graduate students write bland theses in archaic language, proving competency in statistics and established research methods. Professors write articles, judged by their peers, for scholarly journals read mostly by the authors. It is all part of the process of training professionals so narrowly they cannot communicate, except to say, "We are the experts and you don't understand."

Yes, we need training and trained professionals, people focused on vision and mission, not jobs, security, and retirement benefits. We need to invoke the inner spark and brightness in people. "Only in acts of articulate compassion, in rare and hidden moments of communication with nature," wrote Loren Eisley, "does man truly escape his solitary destiny." Every individual feels the urge to manifest his or her compassion. My curriculum for wilderness leadership would help to cultivate and articulate compassion, to show that human destiny need not be solitary, and that through involvement in the social and political process of decision making man's hope becomes man's fate.

We are now paying the price for industrial progress with its overdevelopment and overconsumption, in public health, flirting with our own inevitable Bhopal or Chernobyl or some other toxic disaster. But the most serious effect is in the psychology of people. How sad that we should accept alienation from the earth, that we should even countenance talk of "acceptable risk" in terms of hazardous production materials, or "acceptable change" in terms of wilderness use.

We of our generation can no longer experience the full natural wonders of yesteryear, yet are told the future must "accept" even less of them. I believe the feeling to conserve is deep-rooted in the minds and hearts of people, more than the urge to exploit and degrade. We need leaders to spark the positive. "To do good works is noble," wrote Mark Twain. "To teach others to do good works is nobler, and no trouble." A snowflake by itself may be considered fragile and feeble, but reflect for a moment on the power of snowflakes when they stick together.

Government and, indeed, all institutions are what we want them to be. We get what we demand, which becomes what we deserve. It all begins with thee and me. I see the movement to protect and preserve ever strengthening until it succeeds. It is the ultimate way of life.

Everything Changes with Crowds and Jingling Cash Registers

At National Park Service Tourism and Parks
Conference, Des Moines, Iowa, April 14, 1987.

This may seem bizarre, but the keynote speaker at the 1987 National Park Service conference on tourism in Des Moines was the president, or executive director, of the national association of amusement parks. I listened to him carefully but did not detect any knowledge or appreciation of national parks on his part.

I did observe, however, that the half dozen superintendents of national park units present at the conference were quite attentive to him and to his fellow tourist and attraction entrepreneurs. It struck me that I had rarely, if ever, seen park superintendents as buddy-buddy with citizen conservationists.

Years later, in February 2000, I recounted that episode while speaking at the annual meeting of the Voyageurs Region National Park Association in Minnesota. I was followed on the program by the superintendent of Voyageurs National Park, Barbara West. She spoke earnestly of her desire to leave the park in better natural condition than she had found it. Then she picked up on my remarks. She told of attending a conference of parks and attractions people in Florida similar to the one in Des Moines. She felt uncomfortable and out of place, especially when a spokesman for Disney World reported on the solution to a problem with the display of real-life lions. The problem arose because tourists wanted to see lions move about and behave like bold beasts of the jungle, but the lions were content to lay around at rest all day—until management solved the problem by installing jets of cold or warm air coming up from the ground as needed to energize the lions.

Yes, I've been a tourist, too, and there should be a spectrum of choices. But there is too much sleaze to it, and hyping up the lions is probably the least of it. Mass tourism—based on huge jetliners, huge cruise ships, huge hotels, and charter buses demanding wider and wider highways—has degraded wonderful places in the United States and around the world. Local people can't afford housing because hotels and condos have taken the real estate and sent prices sky-high. They become strangers and servants in their own homeland. In Hawaii in the 1960s, Hawaii residents outnumbered tourists more than two to one; by 2000 tourists outnumbered residents six to one, Native Hawaiians thirty to one. Those natives feel alien in a land of megaresorts with sunbathing, jet skiing, and boozing, where their culture is commercialized and they are expected to work as porters, busboys, housekeepers, and bartenders.

In my travel writing days I connected with wonderful people, many of them still good friends of mine. The more environmentally conscious I became, the more I tried to raise the sights of my readers. I will cite two examples:

In the November 1963 issue of *Woman's Day* I wrote about "Ten Lovely Train Rides." I was trying to make the point that railroads still provided a valid way of getting around, less polluting and less environmentally damaging than private automobiles. I had no way of knowing that a reader in Binghamton, New York, would share this piece with her son, a high school freshman and rail buff. That was my first contact with Alfred Runte, who would grow up to become a historian, university professor, author of books on national parks and railroading, and a close friend. In 2001 he showed me that he was getting his message across in unusual places, including the back cover of the dining car menu used aboard the celebrated train the Empire Builder. His essay on the page began as follows: "One of the great trains of America since 1929, the Empire Builder under Amtrak remains forever linked with Western history. It is a legacy beginning with conservation and the establishment of Glacier National Park, Montana (May 1910). Original park buildings, all railroad-sponsored, include the station at East Glacier Park and magnificent Glacier Lodge."

On another front, in the 1970s I fell in with a fellow named Mike Drakulich when he was running a scuba-diving program at Ocho Rios, on the north shore of Jamaica in the West Indies. I had tried scuba diving elsewhere but was pretty poor at it. However, Drakulich picked me up and took me down. Together we probed the underwater world, vibrant and alive. I found treasures in the coral rising like trees, mountains, and spires; the sea feathers and violet-hued sea fans swaying in the current; anemones carpeting the reef like shrubbery; and the brilliantly colored fish gliding through caves, crevices, and boulders or resting on stony branches. But Drakulich also showed me how onshore construction of roads and development of high-rise hotels were destroying the coral reefs. "It takes nature thousands of years to build these great reefs," he told me, "and only a few short years to kill them forever."

Drakulich initiated a Jamaica reef-protection campaign, and I tried to help by writing about it in the travel pages. I later lost track of him, but he made me reef-conscious and critical of new illusions substituted for old reality in reef-bordered islands of the West Indies and elsewhere in the world. I recall visiting Virgin Islands National Park soon after it was established in 1956. Coral reefs were a particular pride, equated with Yellowstone's geysers and Yosemite's waterfalls. Thirty years later they were shadows of themselves, deteriorated and dying as a consequence of human disturbance and abuse. There isn't much coral along the Trunk Bay underwater nature trail, but divers and snorkelers can still read the labels etched on submerged glass plates describing reef life and imagine they are seeing it.

Early in this book I mentioned the 1959 *New York Times* report by Joseph Wood Krutch on the oncoming tourist boomlet in Baja California. His dispatch was datelined La Paz, where forty years later by chance my wife and I chose to vacation. It was changed since Krutch's day but was still attractive, congenial, and relaxing. One day, however, we took a trip by bus to Cabo San Lucas, the celebrated resort at the tip of the Baja peninsula (one hundred miles south of La Paz). It certainly wasn't the way the brochures show it. The dilapidated bus station and unpaved streets in the slums at the edge of town accent the reality of Mexican life. Downtown Cabo San Lucas came across as a conglomeration of condominiums, construction, and crowds, many Americans in tank tops and tight shorts, and Mexicans hawking their wares and excursions like the Booze Cruise.

In the speech below I quoted a friend, Robert Giersdorf, a successful travel entrepreneur: "Our very existence and success depend on making sure that pristine and unimpaired wilderness experiences are preserved for tomorrow, next week, next year, and for the next generation of visitors to enjoy." I believe he meant it, but there is very little social responsibility in his business.

■ ■ ■ ■

Our national parks undoubtedly are the most popular and most loved tourist destinations in America. That's all to the good. But like any object of beauty, a park requires protection, with high standards of care and conservation, to sustain the qualities that make it special.

National parks should never be regarded simply as tourist attractions with dollar signs attached to them. I see public recreation as a large and essential factor in contributing to the quality of American life. It serves the economy as well, but that isn't its primary purpose.

Outdoor recreation spans a variety of interests, tastes, and goals. Theme parks, such as Six Flags, Busch Gardens, and Disneyland, fill particular niches. So do commercial resorts and campgrounds. But public recreation areas are something else again, filling a different niche.

Public parks and forests provide an antidote to urbanized living, a return to pioneer pathways, a chance to exercise the body and mind in harmony with the great outdoors. In such places Americans learn to understand and to respect the natural environment. Historic parks maintain the opportunity for successive generations to learn firsthand about the conditions that shaped our culture. Contacts of this nature instill the vital sense of being an American.

The national parks assuredly were designed for use by the people, not for an elite aristocracy, nor for scientific study alone. Access to parks is a hallmark of American democracy. But with crowds and jingling cash registers, everything changes. I've seen beauty spots in this country and other parts of the world overexploited and milked dry long before their time. The words "Miami Beach" symbolize uglification of nature for profit. So does the word "Waikiki." Golden arches, chain motels, convention centers, highway strips overcommercialized with billboard blight, and honky-tonk tourist traps—they are look-alikes that blot out distinctiveness and beauty.

Forty years ago Gatlinburg, Tennessee, was a friendly country crossroads and gateway to Great Smoky Mountains National Park. Today the "number one mountain resort of the nation" is like an obstacle course of money machines called "family fun attractions," while the fifteen-story

Sheraton-Gatlinburg Hotel sits perched above the town, flush against the park boundary. There is no escaping the view of this massive white mausoleum, not from hiking trails in the parks, nor from the valley below.

Elsewhere in the country, the most cherished national battlefield parks—Gettysburg, Antietam, Manassas, Fredricksburg, Chattanooga, and Vicksburg—all have been tarnished by commercial attractions, saloons, souvenir stands, subdivisions, and condominiums encroaching from surrounding lands and sprinkled throughout private inholdings within the parks.

The most glaring example of these eyesores is the 307-foot-high commercial tourist tower dominating the scene of Lincoln's Gettysburg Address. The tower distracts the eye so thoroughly that a sense of the tragedy of the battle and of Lincoln's eloquence is lost. This should never happen, considering that national parks are set aside to show the best of America to Americans and to visitors from all parts of the world. [Ultimately this tower proved an embarrassment to Gettysburg and a commercial flop; in 2000 it came down.]

Each park needs a deliberate design to preserve the fundamental values of its natural ecosystem and historic integrity. Otherwise, the inroads of cumulative damage are inevitable. Inch-by-inch losses are accepted because they seem inconsequential, but, as at Gatlinburg, Gettysburg, and a hundred other places, they add up to the loss of values that can never be replaced.

The same holds true of the types of activities sanctioned in national parks. The greatest reward comes from the challenge of doing something on one's own that demands an expenditure of personal energy that yields the feeling of self-sufficiency away from a supercivilized world. Yet helicopters and planes are increasingly used for sightseeing in national parks. The Grand Canyon represents the absolute worst example of furnishing instant wilderness at a price, while jarring the experience of those endeavoring to meet and appreciate the canyon on its own terms.

It grieves me that national park administrators themselves lose sight of their mandate and mission. As I observed on a recent visit to Virgin Islands National Park, the emphasis in management is on serving development and tourism without concern for preservation of the natural ecosystem. I found reefs damaged and dying, a building boom underway where building should never be allowed, and the national park actually contributing to resource degradation by cutting down handsome palm trees to build wide highways and parking lots.

At Olympic National Park in Washington State, the superintendent directed construction of a power line through a wilderness study area, in violation of the Park Service's own policies. When I interviewed him about it, his response was simple: "Do you want to keep the ski lodge from opening on schedule this winter?"—as though that were more important than protecting the Olympic wilderness from the endless nibbling that undermines it.

And now I have at hand a statement from the Voyageurs Region National Park Association, a citizen group, in response to the official proposal for a trail plan alternative. The association objects, very properly, to the portion of the "preferred alternative" providing for a snowmobile trail across the Kabetogama Peninsula, one of the outstanding natural areas of the upper Midwest, furnishing habitat for eagles, wolves, and moose. Snowmobiling has been accepted as a fitting use in the park, but that doesn't mean it needs to be everywhere in the park.

The point is that national parks cannot be all things and still be national parks. Prudent and intelligent people must realize that unrestrained pressure on the parks for profit is not progress. It serves to make one generation rich and to impoverish the future. A place of beauty is like a theater; it may be built to seat five hundred persons. If it is, you don't try to cram one thousand persons into those five hundred seats or to give them free rein to do whatever they want.

I believe the travel industry and citizen conservationist organizations can work together for the long-range good of the parks and the public interest. In Jackson, Wyoming, last year, I learned of a tourism survey commissioned by the chamber of commerce. It determined that visitors were attracted most by the following assets: Grand Teton National Park; Yellowstone National Park; big game, visible and legendary (moose, elk, deer, coyote, bighorn sheep); outdoor recreation, adventuresome and tranquil; mountain setting and scenery; uncrowded open country; and hospitable, friendly people.

Let us work together to keep it that way. The emphasis needs to be on protecting and enhancing the quality and character of each park, and letting dollar values follow. When the desires of business interests for profit are allowed to dominate, the beauty will be lost—inevitably, and without fail.

"Our very existence and success depend on making sure that pristine and unimpaired wilderness experiences are preserved for tomorrow, next week, next year, and for the next generation of visitors to enjoy," states Robert Giersdorf, former president of the Travel Industry Association of America, whose company, Exploration Holidays and Cruises, is a major operator in Alaska and elsewhere in the world.

That makes for a sound approach. It helps to protect a valuable economic resource, rather than squandering and ruining it. At the same time it preserves the parks as the symbols of a national heritage.

Preserving the Endangered Principle

In tribute to Sigurd Olson on his induction into the
Wisconsin Conservation Hall of Fame, Stevens Point,
Wisconsin, April 25, 1987.

I always felt that Sigurd Olson was an exemplary communicator. I mean that he spoke the way he wrote and wrote the way he spoke, with clarity, simplicity, directness, and imagery. He didn't need big words or complicated sentences to make his point. He made an impressive appearance because he was calm and self-assured, never dressed to kill but always neatly dressed.

He was a model for me, and I used the Olson way as a model in teaching. I discouraged students from trying to show their learning with sheer jargon and from showing their individuality by wearing torn, ragged jeans. I told them about my years in Washington, D.C., where activists would come with long hair, shaggy beards, and ragged jeans to testify before congressional committees. Their dress and demeanor came across to conservative congress members as sheer defiance and to progressive members supporting their cause as embarrassment.

When I gave the speech below at Stevens Point, Sigurd Olson's widow, Elizabeth, was in the front row. I didn't want to be mawkish about it but felt privileged that I could say these words as a gift to her. She was the greatest part of Sig's life. She and the work, inseparable.

■　■　■　■

If you offer the people excellence, they will find it out and respond to it. I wish I could claim credit for that saying as original with me. Actually I learned it years ago at Cooperstown, New York, the hometown of James Fenimore Cooper, one of our earliest environmental writers, who through his celebrated fictional character, the Deerslayer, decried the destruction of the forests and the exploitation of the Native American Indian. In more recent times Cooperstown has become the "Village of Museums"—the Farmers Museum, Folk Art Museum, and Baseball Hall of Fame—and it was the benefactor of the entire complex, Stephen C. Clark, who defined his policy as being "If you offer the people excellence, they will find it out and respond to it."

That idea binds together all of those whose work has earned their places in the Wisconsin Conservation Hall of Fame. It offers to us as well a particular challenge for life in our times, an age when excellence—or the effort to attain excellence—is critical, but hard to come by; an age characterized more by conformity; an age, despite all of our industrial and technological advances, threatened by intellectual, cultural, and spiritual mediocrity. The state of things today makes it difficult to pursue a course based on offering excellence with confidence that people will respond to it. And yet the only hope for the future is in doing so.

The honorees in the Wisconsin Conservation Hall of Fame demonstrate that the quest for excellence can and will prevail in the American scheme of things. John Muir, Aldo Leopold, Ernest Swift, Gaylord Nelson, and Sigurd Olson all rose above the norm and by so doing succeeded in raising the norm. That is the message they bring to us and the challenge they convey. I say this especially of Sigurd, on the occasion of his induction. It isn't enough to say, "Aren't they all wonderful?" and to glorify Sigurd. It's not what any of them would want. Rather in their spirit it behooves us to pursue their ideals and ideas and to come away from here with a new determination to rise above our time and thus to make it better.

"Of all God's feathered people that sailed the Wisconsin sky, no other birds seemed to us so wonderful." I am quoting John Muir in describing the great memorable day when the first flock of passenger pigeons arrived at the family farm, living proof of the stories he had read about them while at school in Scotland:

> The beautiful wanderers flew like the winds in flocks of millions from climate to climate in accord with the weather, finding their food—acorns, beechnuts, pinenuts, cranberries, strawberries, huckleberries, juniper berries, hackberries, buckwheat, rice, wheat, oats, corn—in fields and forests thousands of miles apart. I have seen flocks streaming south in the fall so

large that they were flowing over the horizon to horizon in an almost continuous stream all day long, at the
rate of forty or fifty miles an hour, like a mighty river in
the sky, widening, contracting, descending like falls and
cataracts, and rising suddenly here and there in huge
ragged masses like high-plashing spray. How wonderful
the distances they flew in a day—in a year—in a lifetime! They arrived in Wisconsin in the spring just after
the sun had cleared away the snow, and alighted in the
woods to feed on the fallen acorns that they had
missed the previous autumn.

I feel cheated that I can't enjoy the same sight of the skies alive with
birds in flight. And I feel angry. Muir quotes Pokagon, an Indian writer of
his time: "I saw one nesting-place in Wisconsin one hundred miles long
and from three to ten miles wide. Every tree, some of them quite low
and scrubby, had from one to fifty nests on each. Some of the nests overflow from the oaks to the hemlock and pine woods. When the pigeon
hunters attack the breeding-places they sometimes cut the timber from
thousands of acres. Millions are caught in nets with salt or grain for bait,
and schooners, sometime loaded down with the birds, are taken to New
York, where they are sold for a cent apiece."

I feel angry that we still have a price tag on the earth, its resources,
and its creatures, that we still allow the sale of it all at the rate of a cent,
or a buck, apiece; that in the name of jobs and progress anything goes;
that we don't recognize the progress paradigm as a coverup for good
old-fashioned profit and greed. The search for excellence demands that
we plainly identify the difference between the standard of living based
on super consumerism and waste, that makes us enemies of ourselves,
and the quality of life that enables us to live in harmony with our environment and with each other.

I've been conscious of this idea during Earth Week. How fortuitous
that this Hall of Fame induction should come at the end of Earth Week,
celebrated wonderfully here at the University of Wisconsin–Stevens
Point. I remember early in 1970 visiting with Senator Gaylord Nelson at
his office in Washington and hearing him outline his bold plans and
dreams for the original Earth Day. It was a great display of concern and
determination all across the country, convincing me never to lose faith
in our young people. The resurgence of Earth Day 1987 demonstrates

anew that young people, if given the chance, will justify the confidence in them.

Sigurd Olson exemplifies determination and faith—faith in himself, faith in the natural environment, faith in the ability of people to respond to the call for excellence. He didn't become a serious writer until he was fifty; he persevered in the face of adversity, insisting that his writing must be purposeful and elevating. He wasn't a writer only, but a participant, an activist in the crusade for a better world. There are many things we don't know about him. Just last week his old friend Steward Brandborg, the executive director of the Wilderness Society, recalled for me how Sigurd was a driving force in the councils of the society to press on for passage of the Wilderness Act. I hadn't known previously of this role he played, but it doesn't surprise me.

I need to summarize with specific thoughts to match the moment.

First, let us take the spirit of the Wisconsin Conservation Hall of Fame into our lives and encourage it in others. Let us not compromise with excellence. I give speeches and lectures here and there. At one college recently, I talked of my faith in young people responding to the environmental challenge. During the question period an administrator of that college said to me, "But the surveys and statistics show that today's young people are interested in business careers and in making money." My response was that I can't deal in statistics but in people, and I see young men and women who are crying for training as leaders in an ethical world. Moreover, to emphasize business, money making, and serving self in an education curriculum is to acquiesce rather than lead.

I can illustrate this point in another way. Yesterday, at the University of Wisconsin–Green Bay, a graduate student told me of applying for a position as a summer naturalist at a camp for children operated by the Wisconsin Department of Natural Resources. On the application form he pondered one particular question: "Would you approve the use of pesticides?" He wondered whether to play it safe or to answer with his true feeling. He chose to respond that he would *not* approve the use of pesticides, a response that stirred hostility in his interview for the job. "You mean, you would not use a pesticide even if the children were threatened with poison ivy?" "No," my young friend said, "I would much rather explain poison ivy to the children."

He didn't get the job, though he would have made an excellent naturalist, the kind that is most needed. But I choose to honor him for sticking to principle.

Let us honor all those Wisconsinites who place principle above expediency in the great crusade for a better environment for this state and nation. I feel privileged to have known Sigurd Olson. But I would also like to mention Henry S. Reuss, an environmental leader in Congress for two decades; Charles Stoddard, who helped draft the first conservation message of President John F. Kennedy and then sparked the fight to protect the quality of Lake Superior; Hugh Iltis, the botanist, who has stressed the importance of biotic diversity in the administration of public forests; and Richard Thiel, the biologist, who has made us conscious of the wonder of wolves in our world—not in some remote somewhere else, but here.

Few people have ever heard the sound of a howling timber wolf. Only a small number of wolves now live in Wisconsin—where once there were ten thousand to twenty-five thousand—and yet Wisconsin is one of just five states where wolves exist in the wild.

To quote from Dick Thiel, "With so many adult wolves being killed by Wisconsin citizens every year and with few pups surviving to replace them, the outlook for Wisconsin wolves is grim. But steps are being taken to reverse this trend, and each Wisconsin citizen can help."

Let us help. Let us take the message to the people, with hope and confidence the people will respond to it. Let us prove the Wisconsin Conservation Hall of Fame is more than a place, but a purpose. Let us show the spirit of Sigurd Olson is with us. Let us leave a record for the future that ours was a generation that cared.

Wildlife History, with
Ecological Perceptions

At a workshop of Forest Service biologists,
Chelan, Washington, March 7, 1988.

Establishment of forest reserves in 1897 (later renamed national forests) marked the first attempt on a large scale to administer a great national resource in public ownership in the United States. Thus, early in the twentieth century, where there were cowpokes and cattlemen, homesteaders, loggers, and miners, there likely were forest rangers, too. The foresters did not quite represent law on the frontier, but they represented order on the public domain. They included old Indian scouts, rodeo artists, and Spanish-American War veterans.

The emphasis in the early days was on decentralization and grass-roots organization, keeping things simple, open, and close to the people. Some of that atmosphere and spirit was still evident when I first began to observe and write about federal conservation agencies fifty years ago.

The Forest Service began in the West, but Gifford Pinchot pressed to build a national organization. In due course the agency became professionalized and institutionalized. Nevertheless, at least up to World War II and the 1950s, while life in America was still uncomplicated, many rangers functioned free of bureaucratic regimen. Arthur Woody, who began his career in 1912 and worked as a ranger for thirty years in northern Georgia, ignored regulations but successfully surveyed, purchased, and protected a quarter of a million acres. He followed a simple creed: "We should look at forests as a source of good as well as wood." When he died in 1946, fifteen hundred people attended his funeral services at a backcountry church.

A few of those rangers recorded their experiences and feelings. One of them, Harold Criswell, who served thirty-seven years from 1933 to

1971, lives in Bellingham, Washington, where I do. Criswell self-published a memoir, which he made available to me.

He records starting as a junior forester for two thousand dollars a year. There was no such thing as overtime pay, and it never entered anyone's mind. His boss, the district ranger, was a key man in the community: "He ranked as a friend to all and on equal status with the mayor, grocery store owner and owner of the hotel. The job of public relations wasn't something you planned, it was something lived."

Criswell joined the Forest Service in custodial days of number 9 telephone wire strung mile after mile for communications everywhere, of backpacking and horse work, and of lookouts and fire guards—all before radio communication, radar, plastic, ball point pens, credit cards, computers, and television.

His career reached from the tag end of the horse ranger days to the beginning of computers. He was a district ranger on four diverse districts, staff officer on a national forest, and finally supervisor of the Mount Baker National Forest in Washington state. He concluded, "This could be called the golden age for those of us lucky enough to have had a career in the Service during those times." And whenever I speak to the Forest Service I try to rekindle the spirit.

■ ■ ■ ■

I've been asked to address Forest Service origins and history, which I am glad to do. I hope that you share my appreciation of the study of history, and especially the history of public lands. However, I can hardly be content to stop there. If I come to a program of this nature I want to give my critical and hopefully constructive and challenging assessment of your agency's fish and wildlife policies, practices, needs, and goals.

At the outset, I will recall a conversation I had in 1981 with Dale Jones, then director of wildlife and fisheries of the Forest Service. I was spending a year in residence at the Pinchot Institute for Conservation Studies in Pennsylvania, researching Forest Service history and writing about it. Dale telephoned from Washington and drove up to visit. He was plainly concerned about problems of protecting wildlife on the national forests. He spoke of the need within the agency of "ecological perceptions" and responding to "a demand for the peace of mind that comes from protecting and preserving wildlife because its users see it as a treasured gift."

Ecological perceptions and preserving wildlife as a treasured gift ought to be paramount in personal and professional goals of biologists working on public lands. I will ask you today how, and if, we can affirm these in practice.

If we look back far enough, the whole Earth was a wilderness, as though God had created Earth National Park. Then, through centuries and millennia, much of it was used and abused, as evident in the barren hillsides and valleys of China, the Middle East, and the Mediterranean.

The New World, lightly settled and unexplored, remained a national park. With European colonization, however, the forests here too were cut. The more felled or burned, the better; there would always be more—that was the philosophy from colonial days to the late nineteenth century. Following the Civil War, new industries and new cities arose, all utilizing and clamoring for wood. The period was marked not simply by use but by wasteful exploitation and devastating fires. Carl Schurz, the German-born secretary of the interior, in 1875 called for a reversal of public opinion, "looking with indifference on this wanton, barbarous, disgraceful vandalism; a spendthrift people recklessly wasting its heritage, a government careless of its future."

In the face of dissolution of the nation's treasures, Yellowstone was set aside in 1872 as a public trust. With rising scientific and public concern, the foundation was laid not only for additional national parks but also for the Forest Reserve Act of 1891, authorizing the president to withdraw portions of the public domain as forest reserves. President Benjamin Harrison set aside reserves totaling 13 million acres, while his successor, Grover Cleveland, withdrew an additional 20 million acres. Theodore Roosevelt, however, made the greatest contribution, setting aside 132 million acres—very likely his most enduring contribution to the Republic.

The 1897 Organic Act established national forests on a firm footing for the purposes of protecting watersheds and "assuring a continuous supply of timber for the use and necessity of citizens of the United States." In fulfilling this mandate, Gifford Pinchot, the first chief forester, and other leaders of the Forest Service recognized that productive lands and an abundance of resources determine the quality of living of any nation. They were oriented to public welfare, not to private profit. As Pinchot declared, "The earth belongs by right to all its people and not to a minority, insignificant in number but tremendous in wealth and power."

The Forest Service achieved its reputation for square shooting and fearlessness in a system then—as now—constipated with bureaucracy, bungling, and timidity. The Forest Service was different. After Pinchot

was fired by President Taft in 1910, he wrote, "It is the honorable distinction of the Forest Service that it has been more constantly, more violently and more bitterly attacked by the representatives of the special
interests than any other government bureau. These attacks have
increased in violence and bitterness just in proportion as the Service has
offered opposition to predatory wealth."

Those early conservation leaders recognized the role of wildlife.
"The preservation of forests and wildlife go hand in hand," wrote
Theodore Roosevelt. "He who works for one works for the other." But
they didn't know what to do about it. Because the emphasis was on
game animals, they wanted to eradicate mountain lion, wolf, and other
predators. When Aldo Leopold branched from forestry into wildlife, he
learned to appreciate the wolf, but others did not, and have not to this
day. The wolf has largely been eradicated from the national forests, and
coyote control is still practiced and approved.

The wildlife role of national forests differs from that of the national
parks. The latter came into being as reservations to which threatened
species were withdrawn for safekeeping. National forests are open to
hunting presumably as long as the kill is limited to the natural increase
in the game. Nevertheless, national forests complement national parks
where they are adjacent to them and should be administered in coordi-
nation with them.

That is, if ecological perceptions are meant to prevail. But it isn't
that easy. The Forest Service is charged with protecting and enhancing
habitat—food, cover, and water—while states set hunting seasons and
license fees, even on national forests. But state agencies traditionally
concentrate on producing game animals, since they derive most of their
operating revenue from the sale of fishing and hunting licenses and from
excise taxes on guns, ammunition, and fishing gear. Thus the focus of
wildlife management is skewed in that direction, with government programs weighted toward providing for game species, rather than to a
holistic ecosystem.

Timber cutting and game are tied together. Logging has been considered the means of creating more habitat for game species and hunters
have been led to believe they perform a service by "harvesting the sur-
plus." Timber revenue pays the freight for wildlife. Logging in virtually
every instance disrupts the habitat, but pays professionals like you to mit-
igate, or ameliorate, the damage—but not to prevent it. As long as wildlife
is hooked to timber, wildlife in the national forests will continue to suffer.

The Forest Service was assigned major ecological responsibility by the Endangered Species Act of 1973 and subsequent amendments. That law and other laws—the Multiple Use Act of 1960, the Wilderness Act of 1964, and the National Forest Management Act of 1978—all were passed in response to a sense of public urgency and concern. As Senator Hubert Humphrey of Minnesota, principal sponsor of the NFMA, declared, "The days have ended when the forest may be viewed only as trees and trees viewed only as timber. The soil and water, the grasses and shrubs, the fish and the wildlife, and the beauty that is the forest must become integral parts of resource managers' thinking and action."

But I fear those days have not ended for the Forest Service. Considerable good has been done by caring managers, but not nearly enough. National and regional leadership is in a vacuum, almost to the point of honoring mediocrity. Conformity prevails, and with it resistance to ecological principles. As the regional forester for the Intermountain Region, Stan Tixier, said recently, "There are a few species of wildlife that really require a wilderness-type habitat, but not many." No wonder the director of wildlife came to see me in distress.

I'm not giving up, but something must be done to establish confidence in agencies like the Forest Service to manage public lands with responsibility and accountability. Privatization is on the way in prisons, hospitals, public health, and parks, and forests may be next. The challenge to the resource professional in modern society is stewardship—not measured in board feet, game harvest, or visitor numbers, but in safeguarding fragments of the earth, the original America, with the diversities of evolution, explaining their values to society and the consequences of our actions, protecting the rights and options of generations still to come.

You can't do it alone, and neither can I, but together we can make a dent. The Forest Service has a role to play, a role of leadership to recapture. The professionals at all levels in the ranks can make the difference. You can help keep the promise of public service made by Gifford Pinchot by protecting and preserving wildlife as a treasured gift.

We Need a U.S. Wilderness Service

At the Distinguished Visitors Seminar,
University of Washington, College of Forest Resources,
Seattle, April 12, 1988.

I came twice to speak at the University of Washington distinguished seminars. They were meant for graduate students in natural resources, numbering about twenty-five, most of them working for the doctoral degree, and a few for the masters. We had plenty of time for dialogue, all informal, some of it even less formal during the break for refreshments, but there really is never enough time.

Many graduate students have been out in the working world. Several at the U of Washington seminar said they could relate to my criticism of institutional lethargy, based on their professional experience, and welcomed ideas on how to deal with it. I urged them to hang in rather than drop out, but always to stick to principle. Once, after a lecture, a student rose to say, "That sounds very nice, but don't you think it is idealistic?" To which I replied, "I certainly hope so."

I developed a little process in talking to students. Rather than lecture, I tried to open a dialogue so they could hear themselves and their peers. I asked them to take notes while we talked and allowed time before the end of the session for them to write a report on what we had done, and then to turn it in for me to read. It proved revealing and enlightening. For one thing, I found university students, more than a few, who did not know the names of their congressional representatives and barely the name of the president. I had occasion to speak twice in a Milwaukee inner city high school for dropouts, part of an alternative system directed by a friend, Daniel Grego. In response to my assignment for a paper on their goals in life, one very pregnant young woman wrote,

"I want to be the first teenage mother in my family to complete high school." I hope that she made it.

■   ■   ■   ■

I appreciate the opportunity to celebrate wilderness with you and to express, at the outset, appreciation to my good friend, Russell Dickenson, the former director of the National Park Service, who is an affiliate professor in charge of the Distinguished Visitors Seminar series. I'm certain that students and faculty all benefit from his presence here.

I would also like to pay tribute to an old colleague of his in the National Park Service, Herbert Evison, a native of western Washington (though he now lives in the East in retirement). Herb is ninety-five or ninety-six. As a youngster he volunteered to serve as a cook when the legendary Stephen T. Mather made his four-day pack trip through the northern part of Mount Rainier National Park in 1919. While on that trip Mather sparked the idea of a Washington state save-the-trees organization, presently to be known as the Natural Park Association, with young Evison as secretary. It was a go-getting group that convinced the legislature in 1921 to establish a state park system. Subsequently Herbert Evison became executive secretary of the National Conference on State Parks, and, later for many years, an effective and outstanding chief of information of the National Park Service in Washington.

Following my dismissal in late 1974 as conservation editor of *Field & Stream,* a periodical then owned by the Columbia Broadcasting System, I received a perceptive letter from my friend, Evison. He wrote as follows:

> One thing is a cinch: In this day and age when every company is owned by some other company, it is virtually impossible to say ANYTHING of importance without stepping on somebody's toes and irritating the financial nerve, directly or indirectly. Who in hell would suppose that anything said in *Field & Stream* would arouse the instincts of CBS? In conglomerates, each part is bound—is virtually obligated—to scratch the back of all the other parts; I don't know of any single element of the American scene that can more insidiously affect our liberties and our freedom of speech.

It impresses me that if my friend Herb had left out the sentence about *Field & Stream* and CBS, his statement would have been just as valid.

Change the word "company" to "institution" and it's even better, for all organizations, whether private or public, profit-making or eleemosynary, academic, professional, government, or what have you, once they become large and self-perpetuating, inhibit, if not repress, individualism, self-expression, and imagination.

Institutions generally, by their nature, breed conformity and compliance. The larger and older the institution the less vision it expresses or tolerates. This is manifestly evident in the case of wilderness. The Wilderness Act of 1964 and subsequent legislation assign four federal agencies—the Bureau of Land Management, Fish and Wildlife Service, Forest Service, and National Park Service—the specific mandate to protect, perpetuate, and champion wilderness. I can't think of a single one of them that has come close to meeting its obligation and opportunity, except, now and then, with rhetoric and reports.

The National Park Service is no exception. Its personnel may voice concern for ecology as a principle, but scarcely as something practical in critical need of defense. The best defense, at least in my view, is an alert and alarmed public. But national parks personnel are generally inward-oriented and poor communicators. They know the public as visitor numbers, but not as decision makers. Woe unto the parks person who goes to the public with faith or trust in his or her heart. The parks person is a professional, which is how she or he learned to appreciate the values of ecology in theory but conformity and compromise in practice.

A professor once said to me, "What this country needs are ideas bigger and better than money can buy." Professions should be the standard-bearers of ideas bigger and better than money can buy; so I think they were, they must have been, in their early days, all of them, law, medicine, science, religion, journalism, forestry, education. But time has mellowed them all. A professional degree is more of a license to employment than a charge to serve humanity without fear or favor.

Gifford Pinchot strongly pressed the cause of forestry education to train professionals in a social movement. Foresters today are technical people, focused mostly on wood production, trained to see trees as timber rather than as components of a glorious forest enhancing the lives of people, communities, and the nation. A few weeks ago the district ranger of the Mount Baker District came to my class at Western Washington University to discuss the new management plan of the Mount Baker–Snoqualmie National Forest. He did a good job and it was kind of him to come. But the proposed management plan he brought to us would road and log most of the accessible timber in presently roadless areas. It would log extensively in spotted owl habitat and in areas of spiritual importance to Native Americans. It would log twenty-five hundred acres in the mountain hemlock zone as a study to see if anything grows back. It would increase soil erosion to 117,000 tons per year. When he was through I asked myself, What has the Forest Service learned since the big battle twenty years ago over the North Cascades? The best answer I could come up with was that the Forest Service, at least in this district, has learned to smile.

But not a whole heck of a lot more. Twenty years ago I hoped the North Cascades would remain under Forest Service administration because of the old system of dispersed recreation. I urged the agency to give the people what they wanted in stewardship and protection because the people would get it anyway. [North Cascades National Park was established by Congress in 1968, with transfer of national forest lands.] That lesson has still not been learned. The Forest Service is woefully ill-equipped to face the twenty-first-century needs and desires of the American people. One of my friends, who became a high official in the agency, a regional forester, confessed to me that he had received "good training but a poor education." That was in forestry, of course. The broader the training the better the scholar. It's the narrowness, the commodity production focus, of forestry that engenders an anti-wilderness ethos.

I don't see much better coming out of the National Park Service. The history of its administration of the North Cascades is written with sorry chapters the agency would as soon forget. The recently proposed General Management Plan in many ways is not much of a plan at all, being full of internal contradictions, with lip service to the idea of ecosystem protection but without dealing substantively with crucial issues. For example, there are major threats to the integrity of the ecosystem just outside the park boundaries. Many choice landscape sections in the Mount Baker–Snoqualmie National Forest abutting the national park are slated to be cut within the next five to ten years, yet this is not even mentioned.

Why do these public agencies let the public down? Why should officials be afraid to meet their mandates? Gifford Pinchot in his classy little book of 1910, *The Fight for Conservation,* warned as follows:

> The vast possibilities of our great future will become realities only if we make ourselves, in a sense, responsible for that future. The planned and orderly

> development and conservation of our natural resources
> is the first duty of the United States. It is the only form
> of insurance that will certainly protect us against the dis-
> asters that lack of foresight has in the past repeatedly
> brought down nations since passed away.
>
> A nation deprived of liberty may win it, a nation
> divided may reunite, but a nation whose natural
> resources are destroyed must inevitably pay the penalty
> of poverty, degradation and decay.

Pinchot's important message has long since been filed and forgotten. I doubt that it's much taught in school. Students in most academic
programs are bred to be partners of the system, not to challenge it. It's
part of the nature of institutions in our time. Whether the issue be social
justice, peace, public health, poverty, or the environment, all make good
candidates for study, research, statistics, course work, documentation,
literature, and professional careers, while the poor remain impoverished, environmental quality worsens, and our last remaining shreds of
wild, original America are placed in increasing peril.

Martin Luther King Jr. saw three major evils—racism, poverty, and
militarism—and he saw them integrally linked, each one with the other.
I see the attack on the environment as a fourth major evil, also joined
with the others. All of them reveal the urgent need to reorder priorities.

Consider that in the past ten years the population of our prisons has
doubled. We put more people in our prisons than any of the advanced
countries, except for South Africa and the Soviet Union, and we have the
highest crime rate. Prisons are overcrowded, notoriously inhumane. The
poor and uneducated, society's disenfranchised, feel the fury of the justice system, suffering the worst conditions and the longest terms, while
the insiders, like Ivan Boesky, guilty of mammoth crimes of theft and
corruption, get off lightly, serving their short sentences in country-club
prisons.

In the last eight years the proportion of discretionary nonmilitary
spending has been reduced by 8 percent, while military spending has
more than doubled. The U.S.A., compared with all other nations, ranks
thirteenth in infant mortality, ninth in literacy, and first in weapons production. The United States has spent vast sums for security from the
"evil empire," while with a fraction it could have given humanitarian aid
and eliminated the threat of war.

Something is wrong, critically wrong. Martin Luther King spoke of compassion. True compassion, he said, is more than flinging a coin to a beggar; it is not haphazard and superficial. It comes to see that an edifice that produces beggars needs restructuring. A true revolution of values will soon look uneasily on the glaring contrast of poverty and wealth.

Natural resource professionals ought to be in the lead of the revolution of values. Paul Sears, the pioneer ecologist, defined conservation as a point of view, involved with the concept of freedom, human dignity, and the American spirit. The problem is that compassion must be at the root of the revolution of values, while compassion, and emotion, are repressed in the training of natural resource professionals. I recently gave a talk before an environmental conference in Alaska, after which I received a letter from one of the participants. She wrote as follows: "Not once in the ten years I spent studying forestry and land management while getting a Ph.D. did anyone ever speak about ethics. I think that is criminal." It doesn't have to be that way. "Imagination is more important than knowledge," said Albert Einstein. Carl R. Rogers, the psychologist, in pleading for a new kind of human science, cites the fear among graduate students in behavioral sciences of what he calls creative subjective speculation. "They do not recognize that out of such fanciful thinking true science emerges."

For myself, I see imagination and a subjective value system as a force enabling people to rise above sheer facts, which may not be so factual after all. To say it another way, the power of human life is in emotion, in reverence and passion for the earth and its web of life. I didn't think this up, it's an old, ancient idea. But contemporary society is obsessed with facts and figures, and with modern machinery providing access to even more numbers. Alas, the analytical type of thinking of western science has given us the power over nature yet smothered us in ignorance about ourselves as part of it.

The new type of scientific thinking known as the "paradigm shift" stresses consideration of systems as a whole and the interrelation and interdependence of systems at many different levels. But it also recognizes that scientific knowledge can never be more than approximation, not absolute truth. We need new data, valid data, but it takes something more truly to serve society. It's a feeling that counts most, a philosophy, a love of Earth. To identify with life on the green planet in all its forms is to celebrate human hope and human potential.

Wilderness is our great reservoir of hope—of salvation and redemption, to place it on the spiritual and moral plane. That is why the battle to save the wild places has become a major issue before modern society. Billions have been drained out of the public till for unholy military actions in Vietnam and Nicaragua, to sustain the illusion of democratizing the world through force. But saving the wilderness marks a true test of democratic principles along with the effort to extricate America from its destructive global misadventures.

We must face the twenty-first century with new emphasis on human care and concern. I see our lovely country steadily sinking in the quicksand of technological supercivilization and many of our lovely people buying into it without realizing the pain and strain to come. A fair profit may be defensible, but profiteering has skyrocketed at the expense of social and environmental responsibility. The proposal to open the Arctic National Wildlife Refuge to oil exploration and drilling offers a classic case in point. Manifestly it doesn't have a damn thing to do with meeting human needs. Profiteering should never be glorified, nor confused with social service.

We cannot set aside a little bit of wilderness and say, "That much will take care of the soul-side of America." We must rescue everything that still remains wild and recapture a lot more that has been lost, looking to the future of it rather than to its past. In the battle for wilderness there are no enemies. The children of the poor will become rich for what is saved; the children of the rich will be impoverished for what is not saved. It takes considerable courage to stand up against money and the power of politics and institutions. It takes wisdom, or at least knowledge and courage, to work through the system. The pope when he visited the United States said, "We need more than social reformers; we need saints." I would say, "We need more than social reformers; we need revolutionaries—not to commit violent acts but to press society to reorder its priorities." I ask my friends in the Forest Service and in forestry to join in this commitment; for they, like we, will be remembered not by the volume of board feet logged, the miles of road built, the volume of chemicals doused on the earth, the tons of sediment loosened into the stream, but by their positive stewardship of God's green Earth.

"Responsible land stewardship includes the protection of soil, water quality and fish and wildlife habitat. Short-term economics will not override long-term needs of high quality land management." I didn't say that.

F. Dale Robertson, now chief of the Forest Service, said that in 1974 following logging abuse and resource degradation of the Siuslaw National Forest, of which he was then forest supervisor. I wish he had said it lately. Chief Robertson should be the leader of the fight for sound land management. The policy of promoting clear-cutting, reckless and destructive as it is, and distasteful to the American people must end. Clear-cutting must be replaced by conservative, respectful treatment of the land—with production of timber de-prioritized and true multiple use instituted at last.

If Dale Robertson can't handle it, we will find some other way. Our system of government makes that possible. People who care, who care enough, will have their way. The only thing wrong with our democracy is that we don't always utilize it, or test it to the fullest. But when we do, then look out!

Passage of the Wilderness Act of 1964 represents a clear demonstration of people power. Fruition came after eight years of discussion and debate by the Senate and House of Representatives, and after eighteen separate hearings conducted by congressional committees around the country. The bill was rewritten time and again, passed in the Senate, then bottled in the House. The very idea of legitimizing wilderness was aggressively opposed by the timber industry and by the oil, grazing, and mining industries. The National Park Service and Forest Service opposed it, too. The public may own the land, but the administrators prefer to exercise their own prerogative without sharing decision-making authority. But the people, all kinds of people, rallied to the wilderness cause. The very effort surrounding passage makes the Wilderness Act impressive as a statement of national purpose. For it plainly evoked the feeling of countless individuals throughout the country—and likely throughout the world—who would speak for wilderness if given the chance and would say that natural islands within our expanding civilization are essential to the spirit of humankind.

The Wilderness Act and the National Wilderness Preservation System have proven successful, despite the mediocre, lackluster performance of the agencies in charge. Not a single one of them has directed itself to administer wilderness with genuine commitment. Their grades in my book range from a possible C- or D at the highest to F- at the lowest.

All of the agencies have competent, wilderness-conscious personnel, but they are frustrated and unfulfilled by institutional lethargy, antipathy, and unwillingness to meet the mandate of law and people.

These personnel deserve better. So too does the wilderness. The time is at hand, if you ask me, to consider seriously the establishment of a new federal agency, to be known as the United States Wilderness Service. Since we pay officials in government to serve mining, grazing, timber, and associated other commercial interests, why not underwrite a corps of men and women who will prove government responsive to the people's wilderness cause? The Wilderness Service could undertake many missions now unmet. For example, it could prepare and publish a periodic inventory of wilderness now reserved and survey opportunities to protect additional units by all levels of government. It could, in fact, be responsible for a coordinated approach beyond the scope of federal land. It could delineate the diverse values of specific ecological types, show how they can be saved, and report on threats to them. No bureau is doing these kinds of things today.

The Wilderness Act furnished the techniques for preservation of large tracts of federal lands. Some states have developed their own initiatives from that foundation. Now there is need to identify and to provide statutory protection for smaller tracts in urban areas still in relatively untouched state. No bureau is doing this today.

The Wilderness Service would be deeply involved in research covering ecology, history, archaeology, art, and other aspects of esthetics, economics, utilization, and human impact. Federal land-management agencies have conducted some small studies, but they can't yield an ultimate understanding because their approach is too narrow and efforts of their wilderness-oriented personnel are circumscribed. The holistic approach, after all, is not limited to things, processes, and conditions, but is directed at the relationship between them.

The Wilderness Service would not administer land, but would furnish new ideas for better land administration in the twenty-first century. It would help to determine how much human use an area can absorb without impairing its values. Everywhere I go I observe deterioration and degradation for overuse and improper use, a trend that somehow must be reversed.

Still, determining carrying capacity may be the simplest aspect of use. How to impart to visitor, and nonvisitor, the sense of what wilderness is all about comprises a bigger issue. How to utilize wilderness as an educational and inspirational document so that oncoming generations appreciate and respect the natural world may be the fundamental issue for the Wilderness Service, and for the rest of us as well.

"New opinions are always suspected, and usually opposed," wrote John Locke more than three centuries ago, "without any other reason but because they are not already common." Such is the way of institutions. But not of individuals. Only the individualist can succeed, even in our age of stereotypes, for true success comes only from within. When we look at the revolutionary task of reordering priorities, and the sheer power of entrenched, interlocked institutions, the challenge may seem utterly impossible. Yet individuals working together, or even alone, *have* worked miracles. The odds in Selma and Montgomery, Alabama, also looked impossible; and in our involvement in Vietnam and Central America; and in the long fight for the Wilderness Act; and for each and every piece of wild country that has been saved. Each individual must realize the power of his and her own life and never sell it short. Miracles large and small are within reach. We shall overcome.

Replacing Old Illusions with New Realities

At the annual meeting of the
National Association for Interpretation,
Charleston, South Carolina, November 27, 1990.

In the course of making speeches about the environment I got to visit, or revisit, a lot of America. I went to places I had seen as a travel writer, but now I viewed them differently. I still explored and enjoyed the special points of interest but did not have to take notes, or keep files, on hotels, attractions, prices, how to get there, and assorted other touristic details.

In South Carolina, in advance of the interpreters meeting, I went to see the Congaree Swamp, a beautiful and significant natural area which had lately been added to the national park system and thus saved from logging and development. I felt that was great. In Charleston, however, I saw the big new hotel and convention center that local preservationists had opposed, and that I had written against, along with other proposed convention centers (in the *Los Angeles Times* and *Washington Post*). These structures were in the Charleston style, so seemed appropriate, at least superficially, but to me they were out of scale and degrading to the architectural integrity of the old southern city.

A friend once suggested that I write an environmentally focused *Green Guide to America*. I thought about it, but only briefly. I would rather work on replacing old illusions.

■　■　■　■

I feel privileged to keynote this 1990 National Interpreters Workshop and pleased to address our theme, "The Past is Prologue—Our Legacy, Our Future." I can relate to it, having recently completed the manuscript

of a new book, *Regreening the National Parks.* Most of the book, as it
turned out, treats history as the essential foundation for previewing the
national parks of tomorrow and planning intelligently for them.

How can you know where you're going if you don't know where
you've been? That question is basic. Yet it grieves me to continually meet
good, well-intentioned resource professionals inadequately grounded in
the history of their own fields: foresters who know little of Gifford
Pinchot, landscape architects and park people barely acquainted with
Frederick Law Olmsted, wildlifers ignorant of C. Hart Merriam and Ding
Darling, and toxicologists who have never read *Silent Spring.*

Perhaps the inadequacy is no accident. America, after all, is ever the
land of grand illusions, where it's easier to avoid the past, or to look at it
as a pretty picture book, or a gala musical like "Oklahoma," complete
with song, dance, and happy-ending romance. However, as Thomas
Merton warned in *Conjectures of a Guilty Bystander,* a myth is apt to
become a daydream and the daydream an evasion.

Myths rationalize bigotry, exploitation, homelessness, hunger, war,
and the degradation of the environment. For four hundred years the dominant European American policy toward the indigenous peoples of this
country has been one of continuous genocide. And the same for indigenous animals. Even now the grizzly bear is widely regarded as a "savage
killer." Snakes are "slimy," although their skin is actually very dry; the coyote is "cowardly" and the mountain lion "ravenous and craven." The South
was long a mythic paradise all its own, in which benevolent, cultured
planters loved and protected slaves, those innocent, joyful, songful "dark-
ies." Merton wrote that the word "frontier" began as the symbol of adventure and clear-eyed innocence but acquired pathetic overtones in John F.
Kennedy's "new frontier," trying to keep the myth alive rather than recognize that America had become prisoner of the curse.

Modern America is the victim of a syndrome that glorifies greed,
that pervades and weakens government and all our institutions and professions. Little wonder that priorities are lopsided. The world spends
$1.7 million a minute on military forces and equipment, $800 billion per
year. The United States, in particular, has spent vast sums for "security,"
with illegal and immoral acts in other countries, while with a fraction of
that amount it could have given humanitarian aid and eliminated the
threat of war.

Something is out of whack in a country that spends more than a billion dollars for a telescope while failing to care for its hungry; that

cannot help its mentally ill; that crowds its prisons and condemns the imprisoned to defeat, dependence, and despair.

Society needs transformation—a revolution of ideals, a revolution of ideas in all fields, a revolution of ethics to sweep America and the world. We must, for one thing, alter the life-style that makes us enemies of ourselves. That, however, may be the simplest part. Each of us who wants to make a difference must understand more about the history of ideas that dominate the philosophy and policy of society, that dictate our obsession with facts and figures, more about the analytical type of thinking of Western science that provides power over nature while smothering us in ignorance about ourselves as part of it.

I wish I could cite education as the answer, but much of education as we know it is about careers, jobs, success in a materialistic world. It's about elitism, rather than about caring and sharing. It's about facts and figures, cognitive values, rather than about feeling derived from the heart and soul. It's about conformity, being safe in a structured society, rather than about questioning and restructuring society.

Education has become part of the problem instead of the solution. During the recent summer I read an Associated Press report on campus racism. It showed that many schools have established programs to deal with bigotry, but they are for the most part tokens without genuine commitment or understanding behind them. That isn't surprising. Consider that when the world was created it had a certain unity to it. The world was incomplete, to be sure, and changing, even as it is now, but the parts all fit, each part contributing to the advancement of the earth. But today the human components, presumably the most sophisticated of God's creatures, work against each other: Rich against poor, men against women, straights versus gays, old versus young, black against white, Gentile against Jew, Moslem against Hindu, physically and mentally able against the disabled, the educated versus the uneducated.

It shouldn't be that way, considering that we are all sisters and brothers, born of the same Mother Earth, derived from the same living substance manifest in creation of the earth. But education too often is not a unifying influence. It tends to divide individuals and to repress respect for individuality. The straight-A student is the paragon. Straight A means the student is honored and well regarded by parents, while the student who brings home less—because, perhaps, he or she perceives magic in butterflies or beetles—becomes a family problem. The level of talent is externally defined, even though life itself is the essence of talent.

The thought of reversing course may seem unrealistic and intimidating, but as Willis Harmon wrote in *Global Mind Change,* "No economic, political or military power can compare with the power of a change of mind. By deliberately changing their internal image of reality, people are changing the world. First we must be willing to be rid of the poisonous beliefs that have led to the state of affairs as they exist at present."

In this same vein, but more pertinent to the individual, I will cite an interview with Brian Willson, which appeared in the wonderfully illuminating pacifist publication *Fellowship.* You may remember Brian Willson as the nonviolent protester who in September 1987 sat on the tracks to block a train carrying weapons to Central America and lost both legs. In the interview, titled "The Road to Transformation," Brian explains that his disenchantment began as a lieutenant in Vietnam. Later he went to Nicaragua and began networking with people to explore ways of expressing individual conscience. By working to extricate himself from what he considers a complicity of madness, Brian underwent a healing process and intense personal transformation. He concluded:

> Nonviolence is not so much a tactic as a way of experiencing the world within yourself, of understanding the sacred connection with all of life. It's an understanding of how everything is interconnected and how everything is in a state of interrelationship. We are going against our own nature when we start disrespecting all the other parts of life: people, plants, animals, water, sunlight, clouds. I think nonviolence is an attitude and way of life with a spiritual ecological dimension that is aware of how everything is interconnected and resounds honestly to that.

Yes, *spiritus et materia unum:* The antithesis between the material world and the spirit simply does not exist, since the material world is only the content, or reflection, of the spirit. To recognize this idea is the most important discovery of a lifetime, enriching and empowering the individual to do what he or she believes to be right, regardless of consequences. I love the words of John Trudell, the Indian leader, at the 1980 Black Hills Survival Gathering:

> We must go beyond the arrogance of human rights.
> We must understand natural rights, because all the

natural world has a right to exist. The energy and the power of the elements—that is, the sun and the wind and the rain—is the only real power. There is no such thing as military power; there is only military terrorism. There is no such thing as economic power; there is only economic exploitation. That is all it is. We are an extension of the Earth, we are not separate from it. The Earth is spirit and we are an extension of that spirit. We are spirit. We are power.

John's words are rich and challenging, calling those who hope to heal the earth to join with those who hope to heal the souls of humankind to bring something new to a society in distress. They underscore lessons of history, including current history, still to be learned. For example, despite brief periods of concern or support for the native peoples, there has never been a mass movement of non-Indians demanding that our government honor its treaties and grant them basic rights of autonomy and self-determination. The tribal council system was imposed by the United States government in 1930 to replace the traditional leadership that would not sign away their lands to oil and mineral companies.

Likewise, the Alaska Native Settlement Claims Act of 1971 was designed not to benefit natives but to open the North Slope for oil development. Now, under the guise of settling a fabricated Hopi-Navajo land dispute, our government and both tribal councils have been trying to force traditional people from their land to clear the way for coal strip mines and other mineral exploitation. At the heart of the Indians' struggles for their land and their way of life is the understanding that this must be a spiritual struggle.

I urge interpreters to embrace the spiritual struggle and to face the issues of the troubled world. Though nuclear weapons will never force nations to join in recognizing the limits of a fragile Earth, environmental interpreters can lead in pledging allegiance to a green and peaceful planet. I realize that this isn't easy, but the individual with conscience and courage will find the way.

For example, my friend Gilbert Stucker made his first visit to Dinosaur National Monument, along the Colorado–Utah border, in 1953 completely on his own, not primarily as a professional paleontologist but as a citizen preservationist deeply concerned with proposals to construct two dams across the rivers within the national monument. Like many others, he feared that allowing such a project in Dinosaur would place the entire national park system in jeopardy. Stucker encouraged the Park Service to develop the dinosaur quarry as a positive project, displaying and interpreting dinosaur remains while at the same time interesting the public in the surrounding canyons and the threats to them. Presently, as the project got underway, he was offered, and accepted, a temporary appointment as ranger-naturalist. Visitors would ask questions, giving him the chance to lecture to large and small groups. Stucker recalled later:

> I realized full well that when I discussed the proposed dams, I was exceeding my authority. I was supposed to explain the quarry to the visiting public, not ask people to write their congressmen. At one point the park superintendent called me into his office and said, "I just had a telegram from the Secretary of the Interior directing that no Park Service employee is to discuss the threat of dams in Dinosaur National Monument. I know you've been talking against the dams. If you continue, I will have to separate you or discharge you. I have no choice."
>
> But I knew that I must talk against the dams and somehow rode it through until the question was settled—and Dinosaur was saved.

Another example: One morning at Gatlinburg, Tennessee, in 1966, the Tennessee commissioner of conservation testified in support of the dreadful transmountain road across the Great Smoky Mountains as conceived and proposed by the director of the National Park Service. That very afternoon a young state park interpretive naturalist, Mack Prichard, testified as a private individual against the road. Some years later I asked Mack if he considered his independent action fitting and proper. He responded, "No, I don't think it was fitting. It was pretty risky, in fact. I felt if it cost me my job it was worth it—being honest about the fact I thought it was a lousy idea. You do what you have to do sometimes. I thought it was a sorry idea to build another road. The park would be better off without the road it's got. Then you'd have twice as much wilderness."

My friend Alfred Runte, the historian, while working as an interpreter in Yosemite National Park during the summer of 1980, talked to visitors about national park ethics and ideology. He would begin by asking his audience to recognize that national parks are in jeopardy, then add, "What would you be willing to do to see that national parks remain part of the fabric of American society for generations to come? Would you give up some power so that geothermal development would not destroy Old Faithful? Would you use less lighting at home so that strip mines and coal-fired power plants would not be needed in the Southwest?"

For his troubles Runte was directed to a week of "rehabilitation training," if you can imagine that for a scholar and university professor. He took it all in and subsequently delivered his message as he chose. Runte's travail, however, was not over. Brilliant historian and masterful teacher though he may be, he was denied tenure at the University of Washington. But when one door closes another will open, as evidenced by Alfred Runte's successful new book, *Yosemite: The Embattled Wilderness.* He learned that a program, any program, whether in a university or a public setting, without a theme, a message, is pointless. Dispensing information for information's sake is not what the National Park Service, or any interpreter in it, ought to be doing.

This leads me to mention a recent letter from a Huxley College alumnus about his resignation from the National Park Service. "I used to love this job," wrote Chuck. "I can't begin to tell you what a difficult decision it was." But the politics imposed upon professionals working in the national parks got to be too much for him.

I felt disturbed and wrote the regional director. I wrote that I was not really surprised at the departure of a caring and competent person. I told the regional director that I had lately been to Mount Rainier, where the Paradise visitor complex was like a tourist ghetto. One evening I attended an interpretive program. The subject was listed as "Wilderness," but the interpreter knew little if anything about the Wilderness Act of 1964, one of the landmark laws in the history of conservation, or about the National Wilderness Preservation System; the program on "Wilderness" was disappointing and dismal. The next day I heard one seasonal interpreter, whom I know, endeavor to discuss the status of the spotted owl, the symbol of our vestigial Northwest forests. But another seasonal called him aside with a reprimand, "You know very well that our superiors have instructed us to avoid controversy."

I did, in fact, receive a courteous and considerate response to my letter from the regional director: "Many of us agree with you that Chuck is a dedicated and competent employee. He has been recognized by the National Park Service through special awards and as the nominee of the

Region for the Freeman Tilden Award, the highest honor in interpretation. . . . Poorly trained seasonals certainly exist in the NPS, but from my
own observations there are many permanents and seasonals alike who
are passionately dedicated to wilderness and preservation values."

That's the tragedy of it: that caring people feel stressed and
repressed by job restraints that keep them from speaking from the heart.
Some have lost their chance for advancement, or rehire, or getting
hired, by sticking to principle when expediency dictated otherwise. This
must not be.

The time is at hand, now more than ever, to speak openly. Those
who give the most often don't get the recognition—that is true—but
there is no limit to what you can accomplish as long as you don't care
who gets the credit. Let us remember the words of Mark Twain: "To do
good works is noble. To teach others to do good works is nobler, and no
trouble."

Water Is More than a Commodity

At a conference on the Future of
Western Water at Utah State University,
Logan, Utah, September 19, 1991.

At the time of this conference, Wallace Stegner was ailing with a back problem. I knew him but not well enough to call us friends and was surprised at the invitation to substitute for him. I would like to have heard him speak on western water issues or any other subject.

Stegner was born in 1909 in Iowa (the same state as Arthur Carhart and Aldo Leopold) and died in 1993 following an automobile accident. I think of him in company with Sigurd Olson, who is mentioned elsewhere in these pages. Both left indelible legacies in their writing, Stegner in the West, Olson in the canoe country of the North. Both were learned and poised, handsome even as they aged.

Stegner grew up in Salt Lake City. He established the writing program at Stanford University in 1946 and directed it until he retired in 1971. He wrote thirty books, including twelve novels. His best known works include *Beyond the Hundredth Meridian,* his biography of John Wesley Powell; *Angle of Repose,* for which he won the 1971 Pulitzer Prize; *Big Rock Candy Mountain,* and *Where the Bluebird Sings to the Lemonade Springs.* My favorite, however, is *The Uneasy Chair,* Stegner's unalloyed biography of Bernard DeVoto, his close friend. I should also mention *This is Dinosaur: Echo Park Country and Its Magic Rivers,* which he edited and which proved invaluable in the 1950s campaign to keep dams out of the national parks.

■　■　■　■

I feel privileged and challenged at the invitation to substitute for Wallace Stegner, especially since I am here at his suggestion. Wallace Stegner is one of the finest writers and distinguished westerners of our time. I'm sure everyone here joins in wishing for his early and complete recovery.

I have been asked as the keynote speaker to present an overview on broad issues of water in the West from an ethical and historical approach. I certainly hope my remarks help the conference in its goal of defining the public interest, although the best I can do is define my own personal interest.

Through the years I've observed a great deal of the West, of America, and the world beyond. I find all things connected, a tight little globe wired together. I used to say that the nation with the ability to harness natural resources achieves the highest standard of living. Ten or fifteen years ago that was my interpretation of history. I could cite the record in our own country and the living standards we have achieved because of ingenuity in harnessing our resource riches. I might have warned that leadership in the future will depend not only on use and exploitation, but on sound management, proper development and protection of diminished reserves, and that civilizations live or die, rise or fall, on their resources, primarily on water. Now I recognize that nations are interdependent, that global civilization will rise or fall and live or die as one.

I don't view this only in terms of material commodities. When a living species perishes anywhere on Earth the whole world dies a little bit with it, in spirit as well as substance. When mothers in Iraq mourn the loss of their sons, mothers everywhere surely mourn with them. That is sisterhood, powerful influence for brotherhood, for realization that we are all brothers and sisters, children of the same gods. We have allowed ourselves, alas, to be separated and compartmentalized—into rich and poor; young and old; men and women; physically and mentally able and disabled; black, white, red, yellow; Christian, Jew, Muslim, Buddhist, Hindu—finding in caste, class, and color the illusion of protection from others who are different. In higher education professionals learn languages distinct to their callings, whether in law, health care, science, engineering, economics, forestry, wildlife, all very specialized and restrictive. I like to speak of the common language that calls us home, of spiritual dimensions of the natural world that bring people together to recognize difference of appearance but unity of source, and that help to heal and enrich human heart and soul.

Here I am injecting spirituality or ethics, or religion, or whatever you choose to call it, into a forum of experts deliberating pressing issues of public policy and technology. But it's time for it. In the last decade of the twentieth century we live on a planet deeply wounded, troubled by global warming, acid rain, destruction of tropical forests, loss of wildlife habitat, toxic wastes, poisoned air and water, and the pressures of ever-increasing population. It's time to ask new questions in the search for new and better answers, and the best answers come from the spirit; the most powerful human force, emotion, long repressed in our institutions, cries to be free to work its miracles.

> Too often man is ungrateful for what Mother Nature has given him. His science and technology poison the air, cut down the forests, erode the soil, and make sewers of the streams. Is a befouled nature the price we must pay for our modern comforts, our clothes, our means of transportation or even our health, threatened by the industrial pollution of our natural environment? Are they right who say this is the unavoidable 'overhead' of industrialization?

I am quoting from the lead essay in a special issue of the periodical *Soviet Life* titled "Man and Nature" published twenty-five years ago. It revealed grave failings and deep controversy in the Soviet Union's management of its resources. The editors seemed to me to be reaching out in behalf of Russian conservationists and nature lovers to touch friendly hands, and by so doing find support.

I wondered then why the Soviet Union, a socialist nation presumably serving the common good with rational planning and control of production and distribution, should be plagued with problems strikingly similar to our own. My initial thought was that the Soviet government had been forced *(a)* to industrialize in a short period of time a backward agrarian land and *(b)* to concentrate on defending itself in a world of unfriendly nations. But the editors of *Soviet Life* told it otherwise. They defined the cause of failure as "bureaucratic parochialism," otherwise known as "the unreasoned, unjustified decision making of administrative dogmatism," as practiced by the head of one industrial ministry or another trying to forward what he considered his "own" industry, or fulfill his "own" plan and quotas, while committing violence to nature in the process.

Thus factories and cities along the Volga dumped 280 million cubic feet of raw sewage into the river every twenty-four hours. Industrial plants throughout the Soviet Union focused on production, ignoring even the minimum standards designed to protect air and water. The same system applied in all the countries of Central and Eastern Europe under Soviet hegemony. Conditions worsened with time: The Chernobyl disaster symbolized environmental collapse throughout the Soviet bloc. We know now that the recent epochal upheaval, the assertion of people power in throwing off the yoke of oppression, resulted in large degree from pent-up public fury over the degraded and degrading natural environment.

That system failed because it had no checks or balances, no tolerance for differing views, let alone dissent; it suffered institutional inbreeding, denying the means to renew itself. Any system based on deceit and ruled by force only erodes trust, effectiveness, and leadership. It destroys initiative, programming people to conform, to adapt their personalities and goals to established standards of society rather than cultivate their own potential or challenge society and thus contribute to its enrichment.

I find a powerful message in the European experience, a signal of warning to examine critically our own institutions—political, industrial, economic, educational, cultural, all of them—to reassess their validity as we near the turn of another century. In examining the state of water in the West, there certainly is reason for concern.

Water is a limited resource in the West, requiring and deserving husbandry and stewardship. Yet rivers, streams, and lakes are contaminated by chemical runoffs from factories, farms, forests and from spreading, sprawling cities, and are choked with sediment from overgrazing and logging along their banks. The mighty Columbia River and its principal tributary, the Snake, which once supported more than eleven million native Pacific salmon, may soon bear witness to the tragic extinction of wild native salmon, the consequence of decades of dam building.

Drinking water from wells is often contaminated with pesticides. Lake Whatcom, in Bellingham, where I live, instead of receiving protection as the source of community drinking water, is subject to motorboating, plus residential development and logging, all in the watershed. Groundwater supplies in many parts of the West are dropping dangerously to provide for subdivisions, swimming pools, and shopping malls.

Federal water policy subsidizes water for corporate farms in arid areas of California, while California cities complain of drought. Federal lands, administered by the Bureau of Land Management and Forest Service, were set aside to protect watersheds. However, considering the political power of livestock operators and logging operators, plus the narrow vocational training of resource professionals, these lands are administered primarily for commodity production and the watersheds suffer. Along with the decline in water quality, land health declines, yielding to desertification, the historical conqueror of nations.

The awesome poisoning of Kesterson National Wildlife Refuge in the San Joaquin Valley of California must raise serious questions about irrigated agriculture and drainage. The tragedy began in 1980 when the Bureau of Reclamation diverted irrigation drain water containing selenium, arsenic, boron, lead, and other metallic poisons from upstream arid lands into the wildlife refuge. The safe limit for selenium in drinking water is ten parts per billion; at Kesterson the concentration rose to 1,400 parts, and selenium may now be descending into the Central Valley aquifer. When I learn of waterfowl chicks born without eyes or legs, or with brain bulging from malformed skull, I wonder whether I'm looking at the Soviet system of resource use instead of the American. I wonder about the future of Kesterson. Will it suffer the fate of Winnemucca National Wildlife Refuge in Nevada? Rich in wetlands and waterfowl, that refuge was established in 1930, only to be deauthorized thirty years later, when it lay baking in the sun with its source of water, the Truckee River, diverted for irrigation.

Reclamation was well intentioned and served its purpose. The best conservationists were all for it. The Reclamation Act of 1902 created arable land out of desert. When Theodore Roosevelt dedicated the dam named for him in 1911 at the confluence of Tonto Creek and the Salt River, it symbolized America on the march: the highest dam in the world, impounding as much water as the world's previous three largest reservoirs combined.

Reclamation in its day created opportunity and wealth but kept right on going and growing past its time, impelled by wasteful pork barrel politics and what Russians decry as bureaucratic parochialism, while disregarding new understanding of natural systems that call for the protection of rivers, wildlife, and fragile arid lands.

Much the same can be said for the protection of western watersheds—or I should say the failure to protect our watersheds—by the Forest Service, a bureau that has lost its way as a professional agency. The Forest Service leadership is wholly focused on meeting commodity production quotas, showing little responsibility for watershed values, scant appreciation of plants, animals, soils, geology, or human history associated with the areas in their charge.

Part of the problem, to be sure, derives from the directives to public servants from their political superiors. George Bush led us to believe that his administration would be different, that he would reverse the course of his predecessor, Ronald Reagan, the old sagebrush rebel on horseback. I felt cheered when Bush during the 1988 presidential campaign pledged to be an environmental president in the tradition of Theodore Roosevelt. Those words were easy; I can't see any improvement, except in occasional throwaway rhetoric. Bush in 1988 ridiculed Dukakis [Michael Dukakis, governor of Massachusetts, who ran unsuccessfully against Bush for the presidency] about the pollution in Boston Harbor. He made a commitment for "no net loss of wetlands," but he and his staff have since decided to redefine wetlands to remove millions of acres from protection under the Clean Water Act. Bush's energy plan, his highway plan, his move to undermine the Endangered Species Act and its protection of the spotted owl in the Northwest—these all show George Bush as an anti-environmental Ronald Reagan rerun.

Water should be treated as a precious public trusts everywhere, the basis of a sound economy, a healthy society, and especially in the West. "In California, gold is no longer gold; water is." So I read recently in the *Christian Science Monitor.* But it isn't exactly water that counts so much as a way of looking at water, the land, and each other. The central issue as I see it is in the values sought through public policy and defined by it. Resource use embodies both science and philosophy, but the philosophy is more important by far. It must come first, based on recognizing water as more than a commodity but a life force for the earth and all creatures with which we share it.

John Wesley Powell sought to do that a century ago. In his *Report on the Lands of the Arid Region,* he called for orderly planning to fit settlement and irrigation to limits imposed by scarcity of water, and he proposed to prevent monopolization of water through a cooperative approach. Powell's ideas are never out of date. Still, these are very different times. The West is not an isolated, remote province. Nor can the discussion of water be separated from critical social issues. I'm sure it's true that New York, as they say, is ready to explode; very likely Philadelphia and Miami are too. Violence and killing are common fare in urban America. But when I read about a teenage gunman taking over a classroom in Rapid City, South Dakota, and holding children hostage, or about the latest drug raid in beautiful Bellingham, Washington, I realize how dysfunctional things have gotten to be everywhere. The rape of three-year-old children, serial killings, drug abuse, child abuse, gender abuse, ethnic animosities, human degradation and despair seem to be close at hand, wherever one goes, wherever one lives.

A wholesome human environment reflects a wholesome natural environment. We can't have one without the other. American society in its growth has plundered the forests, dissipated the soil, and poisoned the streams. Let that be the past. Now it is time to come of age, to learn from history and build an ethical future. To consciously advance respect for living nature through private and public institutions and the professions is to advance the cause of human dignity.

The time is long overdue to apply the principle of stewardship, real stewardship, to our entire planet, with the United States as the exemplar. That should be the public interest in charting the future of western water.

Little Wonder Managers
Keep Their Heads
Down and Duck

At a conference of wilderness managers,
Portland, Oregon, May 6, 1992.

In September 1985 I joined the Forest Service wilderness management workshop in the Pecos Wilderness of New Mexico, as mentioned in the speech below. For three days we camped, hiked, and talked about wilderness. Then we hiked out to stay at a pleasant lodge for the last night and a final morning session. For bedtime reading I found a little book by Elliott S. Barker in the lobby of the lodge. The book reminded me that Barker was a New Mexico legend, a pioneer guide, rancher, wildlife conservationist, and author, then still alive at the age of 98 (though he died in 1988 at age 101).

The narrative in the book opened in 1908 when Barker was twenty-one, applying for a job as a ranger with the Forest Service. He took the exam, which consisted of surveying, marking timber to be cut, scaling logs, estimating distances by stepping off courses, saddling and unsaddling and packing horses with typical camp equipment (namely bedroll, provisions, cooking utensils and fire-fighting equipment). There were no colleges of forestry in those days, so Elliott didn't need a degree, but, as he was familiar with these activities, it was easy for him to pass the tests. He became an assistant forest ranger and received instructions from his forest supervisor: "Live together, always ride together, do not go anywhere at night, don't even step outdoors without your pistols. Stay away from Old John's Saloon and teach the natives that the Forest Service is their friend."

Some of that atmosphere and spirit was still evident when I first became involved with federal conservation agencies in the mid-twentieth century. It was not unlikely for an employee to be told, "Here's your job. Do it no matter how long it takes." And the employee would carry on as a matter of course, with sense of purpose. Then laws and regulations changed. In compliance with fair labor standards, the employee was directed, "Go home at quitting time." That curbed enthusiasm and sense of purpose. Then came the merit pay system, pitting one employee against another for the reward (which usually went to the one who carefully stayed in good grace with the bosses). In due time the old spirit faded; it became difficult to tell the mediocre from the committed employee.

Nevertheless, over the years I met many caring, conscientious people in the ranks. I remember participating in a Wilderness Stewardship Course for Line Officers conducted at a retreat center in the Wenatchee National Forest in Washington in September 1993. The opening page of the handbook for the course carried a message from Percy Bysshe Shelley:

> Away, away from men and towns,
> To the wild wood and the downs;
> To the silent wilderness
> When the soul need not repress
> Its music

Somebody had to be touched emotionally by those words in choosing them and somebody else in reading them. In the dialogues we shared, many of the Forest Service personnel seriously approached the course objectives to "strengthen understanding of wilderness values and concepts . . . and build commitment for assuming a leadership role." They clearly wanted to do those things and make a difference.

The following year, in June 1994, I was invited to Montana for the National Wilderness Management Training Course for Line Officers. Some line officers (district rangers and above) were bureaucratic, resistant to restrictions required by the Wilderness Act. But as part of the program we broke into small groups for field trips. I went on a two-night backpack in the Bitterroot Wilderness with half a dozen Forest Service and National Park Service personnel. The group included two young women wilderness rangers, the lowest in rank but the best versed by far in what it was all about. In the evening around the campfire, they challenged the chitchat that satisfied others. "Now let's discuss what we've learned about wilderness," they said. And so we did.

Perhaps I can best illustrate the point about the conscientious and caring with the case of Paul Fritz, who was superintendent of Craters of the Moon National Monument in Idaho when it became in 1970 site of the first national park unit added to the National Wilderness Preservation System.

Paul Fritz was a feisty, stocky New Yorker who studied landscape architecture at Utah State University, worked for a time for the Forest Service, then transferred to the National Park Service. In 1966 he was placed in charge of Craters of the Moon, a striking landscape covering 53,545 acres of lava fields studded with cinder cones. The year before his arrival, a wilderness review had been conducted and completed, in conformity with the Wilderness Act. Field personnel initially had considered a wilderness proposal of 42,600 acres, but this was reduced by the agency to 40,800 acres. The Big Cinder Butte area was deliberately excluded. Roger Contor, Fritz's predecessor as superintendent, wanted to extend the road system around it. "Putting everything in wilderness would tie the hands of future managers," Contor later explained. "We felt we should leave the option open for some limited expansion of front country facilities."

All this is documented in a report titled "Administrative History of Craters of the Moon National Monument," written by David Louter, a graduate student at the University of Washington, published in 1992 by the Northwest Region of the National Park Service. Louter records that the Senate acted first, approving the boundary as proposed by the Park Service. But legislation in the House called for including Big Cinder Butte for a total of 43,243 acres—and so it was finally agreed by both houses. Louter concluded:

> The latter legislation represented the influence of Superintendent Paul Fritz. Fritz, who succeeded Contor in the fall of 1966, disagreed with the accepted master plan and wilderness proposal. In particular, Fritz believed after walking the proposed boundary that the planned road addition encircling Big Cinder Butte was a mistake. . . . Beginning in 1967, he gained support from local communities and environmental groups, such as the Sierra Club and Wilderness Society, who had originally expressed their desire to add more lands to the Craters of the Moon wilderness. . . .

The new Craters of the Moon Wilderness, as it was
called, bore the stamp of Superintendent Paul Fritz,
who spoke against his own agency for the protection of
monument resources.

That sort of thing happens rarely, but Paul Fritz proved that it can.
When he retired from the government Fritz brought his professional
expertise to the citizen movement, but was too much of an individualist
to be consumed by it. He was a bachelor, but a family of friends mourned
his death in 2000.

■　■　■　■

I am always surprised to find myself invited to speak at a meeting of professional experts, for I have been known now and then to rock the boat
and cause a little seasickness among the passengers.

For example, several years ago, before he retired from the Forest
Service, Paul Weingart [veteran Forest Service official] asked me to join
a wilderness managers workshop in the Pecos Wilderness in New Mexico. "Why in the world was *he* invited?" a young woman participant
demanded of one of Paul's staff associates. "But why not?" she was asked
in reply. It was all perfectly clear to her: "Because he doesn't agree with
us." That's about the way it goes. As a critic, I'm encouraged to say anything as long it's favorable. But some of those wilderness experts in the
Pecos hadn't camped out in years, as evident in their mistreatment of the
resource, which I proceeded to bring to their attention when called on
to speak.

Another time I was at the National Park Service training center at the
Grand Canyon, for a session on "remote areas management." Yes, that
was what they called it. I listened to the experts use a lot of words to
massage the issues without facing them, until finally I challenged them
to quit piddling with the "remote areas" gobbledygook and square away
with wilderness, unless they were ashamed of their mandate to protect
and preserve the slender fragments of wild America in their trust.

Last fall I was the token environmentalist speaking at the seventy-
fifth anniversary Vail symposium of the National Park Service. It wasn't
that Director James Ridenour or any of his hierarchy was anxious to hear
from me, far from it, but rather that a public member of the planning
committee, Bill Lane, insisted on a little breadth and openness in the
program. The best part of my speech, if you ask me, was in calling on

Lorraine Mintzmeyer, the deposed Rocky Mountain regional director of the National Park Service, to stand and be recognized. Much of the audience applauded and cheered, as professionals should in appreciation of the commitment and courage of one of their own, who was deeply and wrongly wounded.

I also asked Max Peterson, the retired chief of the Forest Service, who was at Vail, to stand and be recognized in behalf of John Mumma, the regional forester who experienced the coercive pressure of organizational culture when he tried to halt abusive overcutting in Montana and northern Idaho. First he and his wife were humiliated by government gumshoes prying into their personal affairs. Then Mumma was dumped unceremoniously and publicly scorned by F. Dale Robertson, the Forest Service chief. Maybe the worst of it came when hardliners started telling me, "Mumma's a biologist, not a forester, you know, and he shouldn't have held the job in the first place."

Little wonder that professionals keep their heads down and duck tough issues. The mischievous meddling by politicians berserk with power is bad enough, but when leadership caves in and turns its back on diligent field personnel doing their very best in the public interest, the principle of public service is in jeopardy, and the country is in trouble.

If you think that I exaggerate let me quote from a recent letter to the editor of an internal National Park Service house organ. This kind of thing can be dangerous; you may recall the incident a couple of years ago when a congressional committee staff didn't like a reference to politics in the Park Service *Courier* and cut off its funding. Here's an excerpt from the letter:

> When was the last time we took a strong stand on major environmental issues such as overgrazing on public lands, irresponsible mineral development, or the failure to add to the nation's Wilderness Preservation System? When was the last time we told the ORV people to take a hike? . . .
>
> While I applaud the concept of "leading by example," what examples are we going to demonstrate? We aren't progressive about recycling, we don't design for energy efficiency or site compatibility, we don't promote the use of alternatives to fossil fuels, we don't do much about getting visitors out of their cars, we favor commercial

interests over private users in such areas as river permit
allocations, we allow snowmobiles and outboard motors
in pristine places such as Voyageurs and Grand Canyon,
we permit development in major resource areas, we
spend more money fighting drugs than ARPA violations,
we urge our superintendents to do more with less when
we should have the courage to tell them to do *less* with
less, and every year we lose ground in the preservation
and protection of cultural resources.

It is not a pretty picture.

No it is not a pretty picture. I believe that wilderness is the heart of
the American ideal, that those choice fragments of Earth still wild, mysterious, and primeval nourish the soul and spirit of the nation and all of
its people. That we, with respect and reverence, have set aside these
special places is known throughout the world; wilderness preservation
as part of our way of life makes a far better and more welcome calling
card to other nations than all the armed might we can muster. Unfortunately, however, I feel deep concern for the future of our National
Wilderness Preservation System, for the federal agencies administering
its components, and for the professional personnel working for them.

Many of the wildest areas, even though remote, are degraded. Vanishing species of wildlife, driven to their last refuge, are jeopardized by
intrusions, extraneous uses, external pressures. The heart of Yellowstone, the oldest national park, the "flagship" of the entire park system,
has been reduced to an urban popcorn playground. As in higher education, where a university loses meaning without its students, a national
park without its wilderness is hollow, an imitation of itself.

Part of the problem, to be sure, derives from the directives to public servants from their political superiors. Politics and politicians have
played a large part in weakening the mission. I've seen a lot of it over the
years, particularly during the long period I lived and worked in Washington, D.C. I saw the abuse of the system and the coercion of career
professionals to forsake their best judgment.

I remember twenty years ago I was in the office of the deputy director of the National Park Service, Clark Stratton. A secretary entered. "The
speaker of the House is on the phone. He wants the director, who is
away. Will you talk with him?" Stratton picked up the phone. It was John
McCormack of Massachusetts.

"Yes sir, Mr. Speaker, what can I do for you?"

"My parish priest got a ticket for speeding on your parkway to Mount Vernon. You can fix it—that's what you can do for me."

That reflects the level of esteem and expectation held by politicians of both parties in dealing with resource agencies and their personnel. I remember another occasion when I was in the office of the director of the Bureau of Land Management, Boyd Rasmussen. He showed me a document he had been ordered to sign by an assistant secretary, Harrison Loesch, whose idea of public service began and ended with taking care of corporate interests (and who later worked for one of the big coal companies). Rasmussen had just refused to approve the document; he had told Loesch, "I don't have to sign it; sign it yourself." I admired and respected Boyd for sticking to his guns, based on professional principle.

In 1981 I was at the Grand Canyon for the dedication as a World Heritage Site by another assistant secretary, G. Ray Arnett, accompanied by the director of the National Park Service, Russell Dickenson. Arnett was overbearing and crude. He treated the director in the most demeaning manner, like a messenger boy. Following the ceremony Arnett helicoptered across the canyon to the North Rim, where his first question to park people was, "What is there to hunt around here?" I know that Russ Dickenson endured personal insult and injury, hoping to protect his own troops in the ranks.

Rules, regulations, laws—all the best plans to manage wilderness become fair game when they stand in the way of political entrepreneurs and their pals.

Conscientious public servants in all parts of the country feel thwarted and frustrated. Lorraine Mintzmeyer testified on how her efforts to prepare a comprehensive "vision plan" for the greater Yellowstone ecosystem were scuttled by political interference from above and how she was transferred for her efforts. It doesn't surprise me. The chambers of commerce in communities like Cody, Wyoming, through their congressional representatives exercise virtual veto power in the administration of Yellowstone National Park.

Here in the Northwest we know too well of the manipulations by Secretaries Lujan of Interior and Madigan of Agriculture (both former congressmen of the party in power) to block implementation of the Jack Ward Thomas report on protection of the spotted owl. We now must deal with Secretary Madigan's scheme to prevent citizen appeal of national forest logging plans. We see the Forest Service, a proud old bureau, on its knees, its current chief, Dale Robertson, and his closest associates reduced to willing messengers, saving their own scalps by doing what they are told by the political appointees above them. Little remains of the spark and purpose once evoked by Gifford Pinchot, the first chief forester.

The best defense clearly is an aware, alert, and involved public. Pinchot understood that. So did Stephen T. Mather, Ding Darling, Howard Zahniser, David Brower—all the leaders who have made a difference, whether in or out of government. "If we fail to make Americans aware of problems facing the national parks, and to involve them in choosing the right solutions to these problems," Russell E. Dickenson wrote when he was director of the National Park Service, "then we are failing in our responsibility as stewards of these public resources."

"The one thing in the world, of value," wrote Emerson, "is the active soul." The active soul responds to the earth as alive, poetic, dramatic, musical. Traditional Indians place a personalized value on the sacred qualities of land. "The earth is part of my body," said Too-Hool-Hool-Zute, a spiritual leader of the Nez Perce.

Deep personal revelations open the heart to feeling and open the mind to articulate compassion that are the root of wilderness preservation. Howard Zahniser, the principal advocate of the Wilderness Act of 1964, was studious, articulate, compassionate. "We are not fighting progress," Zahniser said. "We are making it. We are not dealing with a vanishing wilderness. We are working for a wilderness forever." Theodore Roosevelt said it a little differently: "Aggressive fighting for the right is the noblest sport the world affords."

Activism on behalf of wilderness is the best kind of response, a thoroughly patriotic response worthy of Roosevelt. A wholesome natural environment provides the foundation for a wholesome human environment. We can't have one without the other. To consciously advance respect for living nature is to advance the national welfare and human dignity. This technical session, in addressing key issues faced by resource managers struggling to protect American wilderness, is patriotism too. More power to you.

Objectivity Is Not the Answer— but Advocacy Journalism Makes the Difference

At the annual convention of the Association for Education in Journalism and Mass Communication, Montreal, August 6, 1992.

When I talk about "advocacy" and "emotion" to journalism educators, they are likely to strike back at me with "objectivity" and "science." But, of course, most of their students are headed for the mainstream media, which provides plenty of information, packaged as facts, but mostly bland and unquestioning. And that information is largely derived from public relations materials provided by public agencies and private corporations serving their own particular interests.

In the foreword to an edition of *Silent Spring* issued after it became a best seller, and after Carson's death, Paul Brooks, her venerable editor, wrote, "Rachel Carson was a realistic, well-trained scientist who possessed the insight and sensitivity of a poet. She had an emotional response to nature for which she did not apologize. The more she learned, the greater grew what she termed 'the sense of wonder.'"

I don't see much sense of wonder manifest in either mainstream journalism or mainstream science. Some years ago I read and kept an article on "The Different Worlds of Scientists and Reporters" in the August 1985 *Journal of Forestry*. According to the authors, G. I. Baskerville and E. L. Brown, "The reporter who wants to survive can ill afford to spend time searching scientific literature in its appropriate scientific background. Few reporters ever read a single paper in a scientific journal. Their sources are more likely to be previous media reports or telephone interviews, simply because they are accessible, fast and convenient."

That is essentially so, but the authors also show a cloudy side of science:

> The environment in which the scientist works affects the way information is transmitted. Advancement in terms of salary and budgets is based on the quantity and quality of research produced. . . . Researchers competing for money must submit proposals to granting agencies who, then, have proposals reviewed by a panel of scientific peers to judge which proposals are most worthy of funding. . . . Researchers are continually aware of peer evaluation of what they do, what they write, and what they say. This has a crucial effect on the information provided the public.

Then I purchased *A Field Guide for Science Writers,* a whole book featuring essays by thirty-plus members of the National Association of Science Writers. Basically it shows science writing heavily influenced and employed by corporations, trade associations, educational institutions, and government agencies. Apparently there's plenty of work and money in preparing press releases, news and feature stories for magazines, speeches, radio and television reports, and research magazines, newsletters, newspapers, and brochures. The federal government alone spends $40 billion a year on civilian science and technology programs, plus another $30 billion on military research and development. That is why many of these agencies hire science writers—as full-time employees and free-lancers on assignment—to get out the message and convince the taxpaying public to keep that money coming.

The trouble is that allegiance of the science-writers-for-hire belongs to the institution that employs them. Or, as Rick E. Borchelt, special assistant for public affairs at the White House in the Clinton administration, a member of the board of the National Association of Science Writers, explains in the *Field Guide:* "The first rule in coping with agency editing is never to get personally interested in anything you write. Nine times out of ten you may not recognize it when it comes out of review."

That is why I came to advocacy journalism, for serious students who want to make a difference.

■　■　■　■

I teach advocacy journalism. I teach it for serious students who come to me with concern for the environment, who want to make a difference in their own lives and in the life of the planet. At the outset, to avoid any misunderstanding, I explain my view that "objectivity" is not the answer, considering it is never, ever, objective. I assure them that advocacy journalism *is* journalism, and no place for propaganda. It is the most honest journalism, and most enriching for the journalist.

Environmental advocacy journalism requires an understanding of the nature and purpose of mass communication; an ability to research and to report findings with accuracy and fairness, and a love of language that facilitates clarity in expression. But they need to know much more than "how to write." I want them to learn the power of emotion and imagery, to expand their awareness—to think not simply of who, what, when, where, and why—but to think whole, with breadth and perspective.

I cite the words of Rachel Carson, who put it quite simply: "Do your homework. Speak good English. And care a lot." I encourage students to write with purpose, courage, feeling, and personal involvement, as Rachel did, and to express a strong point of view, critical when need be, but always based on thorough research and sound data. `

Advocacy writing aims high. "After a few thousand words from her, the world took a new direction," an editor wrote of Rachel Carson in tribute to her classic work, *Silent Spring.* Yes, the works of professional writers, and writing by nonprofessionals as well, have been significant in changing attitudes and policies. In Rachel Carson's case, one lone woman, committed to her research and to clear expression of her findings, made a difference in world thinking. I want students to make their contributions in rural communities, cities and states, and by so doing contribute to improving life on the planet.

The environment needs the media. Neither public nor government will respond to the greenhouse effect, toxic pollution, rain forest destruction, or any other issue, whether global, national, or local, until it is properly reported and interpreted. The best of data, the best of programs ultimately depends on public understanding, discussion, and decision making. Effective communication enables citizens to participate intelligently in this process.

Much of traditional writing, like much of traditional scientific research, tends to be narrowly focused, impersonal, free of "value judgments" that might impair scrupulous objectivity, free of imagination and sense of person. As John Muir once noted, however, "Dry words and dry facts will not fire hearts. In drying plants, botanists often dry themselves." Perhaps the chronic dryness is no accident. Institutions and professions of our time tend to be conformist and conservative, breeding young talent to play it safe, careful not to rock the boat. The media ought to be different, the one institution of society that watchdogs all the others and keeps them honest.

The late Edward J. Meeman, Scripps-Howard editor and pioneer in conservation journalism, expressed this idea in a speech to the North American Wildlife Conference of 1965. "Democracy requires leadership and this must come from newspapers among others," Meeman said. "This newspaper leadership should appear in the editorial columns, but it should also appear in purposive, although fair and objective, studies and exposures in major articles or series of articles exploring situations in depth and detail by writers who are not only reporters, but digging investigators and scientific interpreters." Yes, democracy to function fully and effectively requires courageous leadership, openness and dialogue for citizens to independently shape intelligent opinions. The mainstream media provide plenty of information, packaged as facts, more than anyone can absorb or digest, but mostly bland, shallow, unquestioning; anything beginning with interpretation is suspect. The irreverent, independent reporter is apt to pay the price of a complaint to the editor for that unforgivable sin, "losing objectivity." Thus, strong voices are denied the opportunity to furnish important data and the democratic process of decision making suffers in consequence.

For example, Steve Stuebner's editors at the *Boise Statesman* discovered last year that he was undermining the sacred objectivity of their news columns by injecting what they considered his own personal pro-environment bias. They caught him doing this, for example, by referring to a contested Idaho roadless area as a "wildlife sanctuary." So the editors of the *Statesman,* a unit of the Gannett chain, announced they were shifting him from the environment beat, after five years with a string of awards, to cover city-county news. Stuebner conceded ruffling feathers now and then, but felt victimized by the management's demand for blandness—"milquetoast reporting," as he calls it—and he quit.

Stuebner isn't the first to go down, then out the door, for reporting and writing about the environment as though it really matters. Richard Manning, who worked for the *Missoulian,* the daily newspaper in Missoula, Montana, attracted wide attention in 1988 with a hard-hitting series on the exploitation of Montana forests. It was no small venture, involving weeks of interviews, backcountry legwork, and painstaking examination of documents. He won awards for investigative reporting, but his bosses at the paper felt squeamish. They squelched him by transferring him to another beat, and he quit.

It happens to the best of people in the best of places. Philip Shabecoff for thirty-two years worked as a reporter for the *New York Times.* For ten years he covered the environmental beat in Washington, a pacesetter in his profession, turning up one major story after another. While Shabecoff was respected and admired on the outside, he was rowing upstream on the inside. Editors told him he was "ahead of the curve," that he was stale, biased, "too close to environmentalists"—exactly the same message, in almost the same words, that Manning and Stuebner heard from their bosses. When he was taken off his beat and switched to cover the Internal Revenue Service, Shabecoff quit.

Dan Sholley, the chief ranger of Yellowstone National Park, decries the coverage of the 1988 Yellowstone fires, when the park was invaded by "headline-hungry journalists who came in droves and demanded answers to mindless questions and then sped away to cover in six inches of print, or one minute of television a subject that had taken a whole day to explain properly." That type of reporting can easily be called Wizard of Oz journalism—no heart, no brain, and no courage.

The truth is that big environmental stories are scarcely broken by mainstream media. Most of those stories appear first in alternative publications, like *High Country News* and *Environment Hawaii* that lately have emerged because the mainstream has failed. "We do have a beat, the environment," as Betsy Marston, editor of the feisty *High Country News,* puts it, "and we do care about the land, the animals and plants that live on and with the land, and the communities that have a stake in the West. We think of ourselves as the grassroots, on-the-ground newspaper. Our job goes beyond reporting to interpretation and understanding." That make sense to me as journalism playing its rightful role, and while the big papers complain about loss of circulation, advertising and influence, the semiweekly *High Country News* keeps growing. It is published by a nonprofit educational foundation that doesn't depend on advertising, so it doesn't waste paper. Society needs such publications—more of them in every part of the country—to tell it straight and fire hearts, going beyond reporting to interpretation and understanding. With dramatic advances in desktop publishing and in audio-visual communication, I foresee vast new horizons and opportunities opening to us.

Plenty of good journalists are ready, like those I mentioned above who would rather quit than cave in. They follow the best tradition of their profession, my profession, as practiced by Lincoln Steffens, Upton Sinclair, and the other muckrakers early in this century. "A newspaper is indeed like a woman or a politician," wrote Steffens in his celebrated autobiography. "When it is young, honest, and full of ideals, it is attractive, trusted, and full of the possibilities of power. Powerful men see this, see its uses, and so seek to possess it. And some of them do get and keep it, and they use, abuse, and finally ruin it." No wonder he and others did most of their writing for magazines and alternative papers.

Steve Stuebner calls his old bosses "Gannettoids" and their products "McPaper." Maybe so, but the media has a powerful role to play beyond its current meager efforts. It's more difficult for editors to ask, What's really going on? What does it mean? Where's the rest of the story? And even more difficult to come up with hard answers, as a handful do at the risk of their careers.

Considering that we teach more by what we do than by what we say, students tend to imitate their instructors, often finding their example a good way of staying on the sidelines. However, I believe in self-expression and social involvement as important learning elements. So did Aldo Leopold, the pioneer educator in wildlife science, who was neither bland nor impersonal. He was an advocate, even a crusader, for wilderness and the land ethic. In his teaching at the University of Wisconsin and in A *Sand County Almanac* he demonstrated that emotion and aesthetic sensitivity can be wholly compatible with good science. That is the way it should be, with journalism as well as science.

Education for the most part is about careers, jobs, making it in a structured society, rather than about individualism, the ability to question society and thus to change it for the better. It provides a practical means of acquiring information, but the intuitive, ethical, and spiritual are largely omitted, or denied.

Consequently institutions and professions, directed by highly educated individuals, repress emotion and imagination, though these qualities open the heart to feeling and open the mind to articulate expression. At times, to be sure, an open expression of ideas may seem foolhardy or risky. It endangers professional acceptance and advancement. But freedom of the individual, with the right of self-expression, is sacred. The challenge is to make the most of the democratic American system. It's the way to keep professionals from being weakened by inbreeding, from suffering narrow vision, resistance to change, emphasis on structure, and a dwindling sense of humanity.

That is why in my courses I supplement textbooks on journalism and writing with the works of Paul Brooks, Rachel Carson, Bernard DeVoto, Joseph Wood Krutch, Aldo Leopold, John Muir, Sigurd Olson, Wallace Stegner, and others in the library of environmental literature. I offer citations from Thomas Merton such as the following (found in *Raids on the Unspeakable*): "News becomes merely a new noise in the mind, briefly replacing the noise that went before it and yielding to the noise that comes after it, so that eventually everything blends into the same monotonous and meaningless rumor. News? There is so much news that there is no room left for the true tidings, the 'Good News,' THE GREAT JOY. Hence The Great Joy is announced, after all, in silences, loneliness and darkness."

I endeavor to challenge students to find the true tidings, and to prove Merton wrong by evoking hope, The Great Joy, in light and openness. I allow each to develop his or her own styles, whether strictly reportorial, interpretive, or provocative. That isn't nearly as important as recognizing the spirit of ethical caring that draws them to advocacy environmental writing in the first place, a sense of giving rather than taking, of daring to stand alone when need be, powered by the idea that the great use of a life is to create something that outlasts it.

Media and the Environment: Passion, Politics, and Empowerment

At the Western Regional Conference of the
Association for Education in Journalism and Mass
Communication, Reno, Nevada, April 9, 1994.

I spoke as part of a panel at the Reno conference, together with Steve Stuebner, formerly with the *Boise Statesman,* and Kathy Durbin, then still with the *Portland Oregonian.* In the preceding speech ("Objectivity Is Not the Answer") I referred to Stuebner's removal from the environmental beat and his resignation from the *Statesman.* Over lunch following the program at Reno, Durbin shared the ill tidings that she was being dumped and her days at the *Oregonian* were numbered.

Stuebner and Durbin were highly competent in their work. They covered the environment as reporters, not as advocates, as I might want them to. But they both got in trouble over the way they treated forest issues. Stuebner had committed a mortal sin by referring to a national forest roadless area as a "wildlife sanctuary." The timber industry did not like that at all. Durbin, with her colleague Paul Koberstein, had lately completed a major series showing the realities of Oregon's forest future with reference to water, wildlife, soils, and recreation, as well as timber. Readers welcomed and appreciated the comprehensive scope of their work, but the timber industry did not. Koberstein felt the hot breath of discontent, vented via the publisher, and quit to start an alternative monthly, *Cascadia Times,* and now Durbin's number was up.

Representatives of the timber industry complained about Stuebner in Idaho and about Durbin/Koberstein in Oregon. And the publishers heeded. As it happens, both the *Oregonian* and *Statesman* are part of the same chain, Gannett, which keeps profits high at the cost of editorial quality. Some of its profits go to the Freedom Forum, which espouses the cause of free expression, except for its own reporters.

Stuebner, Durbin, and Koberstein have all survived in their profession. They know that truth telling will prevail, that when one door closes, another opens. Good journalists don't have to go to the mainstream to get their stories published and find fulfillment. All the national environmental organizations have their own communications programs in print and on video, radio, and the Internet. Their periodicals are well designed and edited, with press runs up to half a million copies.

Many other periodicals serve specific environmental audiences, like *American Rivers; Animals' Agenda; Cultural Survival* ("on the rights of indigenous peoples and ethnic minorities"); *Greenpeace Quarterly; Journal of Pesticide Reform; Simple Living: The Journal of Voluntary Simplicity; ZPG Reporter* (published by Zero Population Growth); and scores of regional and community-based periodicals like *Bloodroot,* "dedicated to protection of the heartland hardwood forest"; *Accent,* voice of the Western North Carolina Alliance; and *Raven Call,* of the Southeast Alaska Conservation Council. They all need trained journalists and do better when they have them.

Important environmental stories often appear first in independent, alternative news weeklies and monthlies. Even while newspapers have folded and declined, new breeds of media are filling the gaps of social concern and critical commentary. More than one hundred alternative news weeklies are published in large and small cities. Though some are heavy on life-style, entertainment, and personal ads, others are courageous in investigative coverage of the environment, poverty, racism, downtown decay, and criminal justice; they carry stories in-depth, consider environmentalists as responsible sources, and dare to redefine what is locally newsworthy

■　■　■　■

When my own students and other bright young people ask for advice on how to make a career writing about the natural world, I try to answer from personal experience. I can see they enjoy the outdoors, care about the future of the planet, and want to make a difference in society and their own lives. But experience has taught me there is no magic formula and no magic either. Almost everything I've done came the hard way. The difference emerged when I chose as a writer not to be an observer only but an advocate and participant, involved with people, groups, and live issues. The more involved, the more meaningful I felt my writing became.

Writing for me has been a means, rather than an end. Years ago I discovered wilderness as a cause. I read everything I could about wilderness. I met many of the authors who helped make the Wilderness Act happen in 1964, including Paul Brooks, Bernard DeVoto, William O. Douglas, Margaret Murie, John Oakes, Wallace Stegner, and Sigurd Olson. They were people I respected and admired who had committed themselves and were willing to take on tough issues of principle and politics.

It's been that way ever since. My work in wilderness shows more questions to answer than have yet been asked and much more still to learn. It takes unending curiosity and total commitment, plus willingness to risk and sacrifice, which likely is true of anything worth doing, particularly in a social cause. To quote W. E. B. Du Bois, "If there is no struggle, there is no progress; power concedes nothing without demand." Environmental writing gives voice to struggle and demand—at least that's how I see it.

It isn't easy, there's a price to pay. In tackling critical issues I stepped on the wrong toes. In 1971 my columns in *American Forests* about clearcutting and forest mismanagement caused a problem. The executive director of the American Forestry Association ordered that I was "not to write critically about the U.S. Forest Service, the forest industry, the profession, or about controversial forestry issues." I couldn't accept the censorship and was fired. Dismissal from that magazine taught me that I must continue to call the shots as I see them.

Then in 1974, after seven years as conservation editor and columnist, I was fired from *Field & Stream.* A new editor of the time directed that I write in generalities, instead of holding institutions and individuals responsible for deeds and misdeeds. When I was dismissed, many friends and organizations protested. "We tout our system as being one of freedom for the individual, and we particularly tout our freedom of speech and expression," wrote Tom Bell in *High Country News.* "His [Frome's] experiences give the lie to just how much freedom we really do have."

That was not my feeling, however. Truth telling must and will prevail. Too many writers censor their own work, anticipate trouble, and sanitize their writing to the point of banality. I can't do that. I believe there will always be outlets. When one door closes, another opens. The next outlet may not be as significant as the last, but one never knows where it may lead. After *Field & Stream,* I became a columnist for *Defenders of*

Wildlife, which went on for eighteen years. Life has shown me the world is full of miracles.

Journalism as taught and practiced in the "mainstream" doesn't work. It represses emotion, imagination, and creative expression. Emotion and caring are important to me in writing and teaching. To write something meaningful about the natural world requires vision, caring, courage, and hope. Any writer with those qualities will have a creative, rewarding career, maybe not wholly profitable, but productive and positive all the way. Bernard DeVoto felt the same way. He made this clear in his column in *Harper's:*

> My job is to write about anything in American life
> that may interest me, but it is also to arrive at judgments
> under my own steam. . . . With some judgments that is
> the end of the line; express them and you have nothing
> more to do. But there are also judgments that require
> you to commit yourself, to stick your neck out. Express-
> ing them in print obliges you to go on to advocacy. They
> get home to people's beliefs and feelings about impor-
> tant things, and that makes them inflammable.

We need people who care enough to stick their necks out. In the last decade of the twentieth century we live on a planet deeply wounded and physically degraded. At the same time, the world is divided between those who do not have enough, including many hungry and hopeless, and those who have more than enough and still hunger for more. Violence, killing, brutality, and injustice are common fare, against the earth and against the human family.

I can't *teach* spiritual ecology, but I can tell students about it and allow them to deal with it as they wish. Since my courses deal with journalism and the environment, I do talk about the life and work of Dorothy Day, the founding editor of the *Catholic Worker.* Dorothy saw human life as a procession of large and small miracles; for her it was only a matter of knowing where and how to look: "We would be contributing to the misery of the world," she wrote, "if we failed to rejoice in the sun, the moon, and the stars, in the rivers which surround this island in which we live, in the cool breezes of the bay."

It was her wilderness, as meaningful as a million acres in Montana. I talk about Dorothy Day and about Thomas Merton, and his way of looking at news. News as provided to us, Father Merton felt, was merely a new noise in the mind, briefly replacing the news that went before it and yielding to the noise that comes after it, so that eventually everything blends into the same monotonous rumor. He felt the media was putting out so much news there was no room left for the true tidings, the Good News.

In my classes I see students who want to find and feel the Good News and to spread it. They reinforce a belief that we *can* make it happen, that trying, together, is worthy and rewarding.

Those students need to feel the power in their lives, power to join in determining public policy and the course of history. With power comes a new awareness of human rights, of political and personal freedom, new meaning to the words of Theodore Roosevelt: "Aggressive fighting for the right is the noblest sport the world affords," and, perhaps more appropriate to this discussion, the words of Rachel Carson: "The beauty of the living world I was trying to save has always been uppermost in my mind—that, and anger at the senseless brutish things that were being done. I have felt bound by a solemn obligation to do what I could—if I didn't at least try I could never again be happy in nature."

Rachel Carson rose above the limits of her education to pursue the power of service. Through *Silent Spring* and her resolute personality, she showed that a single individual can make a difference in society. Individuals do that, rising above themselves, and above institutions, without hope or design for material reward, to challenge an entrenched system and make it respond. Over the years I've observed heroic efforts, many, many of them led by individuals timid at first but emboldened by commitment and belief to give voice to the voiceless.

In a thoughtful article in the *Chronicle of Higher Education* (May 29, 1991), Steve Weinberg, of the University of Missouri, wrote that academics with doctorates in journalism tend to suffer in their teaching from lack of passion and consequently are too likely to turn out passionless journalists. He's right, but it doesn't have to be that way; I want to legitimize subjectivity, expression of personal viewpoint in learning as well as in journalism, as a significant means of changing society.

Passion leads to empowerment, as a process from within. Nobody made Rachel Carson but Rachel herself, and the same for Martin Luther King. They felt bound by solemn obligations to do what they could—if they didn't at least try they could never be happy again. Once empowered internally, however, nothing could stop them from externalizing their powerful influence in society.

Reverence for life is a universal human need, expressed in various ways through the movements for peace, for social justice, for equal rights, for a decent, wholesome environment. Heal the soul and heal the earth, I would say. Heal the earth, heal the soul. The message calls for its messengers to prove there is room indeed for the true tidings.

Media as the Messenger

I've been speaking critically of the media, but some of my best friends, as the saying goes, are journalists. Instead of a speech from me, in this chapter I present what they have to say. Reporters came to cover my speeches and sometimes they would interview me. Usually they asked intelligent questions and got things right. They weren't all friendly or sympathetic, but most were. Occasionally they wanted to visit, over coffee or a drink. I think they wanted to quote me saying things they couldn't quite say themselves.

A few have been good friends. I met Sam Venable Jr. when he started as a young reporter in Chattanooga. For years now he's been writing a feature column in the Knoxville paper and putting the columns together every year or two into books. I met Dale Burk when he was outdoors editor in Missoula and digging into the truth about clearcutting in the national forests of western Montana. Dale earned a Nieman Fellowship at Harvard but felt disillusioned when he returned to find changes at the newspaper and in the parent chain. He went into other work, a loss all around.

A few of the reports I saved follow below. I arranged them chronologically, except for the last one, which sums it up for me. I never met the writer and didn't even know he was there.

KANSAS CITY STAR, OCTOBER 1, 1970
Prairie Meeting Fires Tallgrass Hopes
By Ray Heady

Elmdale, Kansas—For the last several years the Tallgrass Prairie Conference has coincided with the annual migration of Monarch butterflies.

The Flint Hills are spectacular enough in their own beauty, but with waves of Monarchs crossing them they take on a new dimension. It's a land of wind, grass and sky, with a shimmering layer of black and orange wings in between.

Like the thousands of migrating Monarchs which swept the Flint Hills last weekend, nature lovers winged into Camp Wood from all directions. The butterflies kept on fluttering south but the 200 men, women and children from nine states paused to attend the fifth Tallgrass conference, an annual event to advance the creation of a National Tallgrass Prairie Park in Kansas.

The weather was ideal. The recent heat wave which had turned the Flint Hill grasses brown and sere had broken. Nights were chilly, mornings cool and afternoons warm. The sunrise bird hikes, canoe trips and photography sessions were spectacular, as were the flaming sunsets.

Ten speakers of national and state stature addressed the conference. They arrived at no consensus, which was expected, but they received intense attention.

Michael Frome of Washington did not arrive like a fluttering butterfly; he came in like a prairie falcon with all talons bared. Frome is an author and columnist for the *Los Angeles Times* and a perennial gadfly to the National Park Service. He has published his National Park directories [annual guidebooks] and his relationship goes back, he said, to the days when the National Park Service was an aggressive department that had "vigor and a certain mystique." But the Park Service has lost that mystique, Frome said, and is now a tame critter with a record that ranges from disappointing to dismal.

"The United States should have four, not one, grassland parks," Frome said—one each in Kansas, the Dakotas, Texas and Oklahoma. The Park Service knows this but they drag their feet. The Service won't act any more. It's up to the people to raise the banner now.

"What we need is a sense of crisis. We don't need another study. The park plan has been studied to death. We don't need a proposed park; we need the park now—1979 at the latest."

Frome issued a warning. "Splitting the park effort by saying Indiana Dunes people want a park and so do the Kansas Prairie people, but we can't have both is an old tactic of the Park Service," he said. "But as I. F. Stone said, government is run by liars. They won't tell the truth. This ranges from the bureaucrats to the Cartercrats. They buckle under when the pressure comes.

"The best way to save tax dollars is to build the park now, not after prices continue to rise. Everyone knows this except the bureaucrats.

"But if the park is to be built it's up to the people at this conference to get the job done. Go to work at the grass roots. Knock on every door in Cottonwood Falls. You will find more friends than foes. Put national organizations to work but don't forget to go to the ranchers and talk to them.

"People power, not the professionals, will build this park. Go for the biggest and best park attainable and keep going. That's the only way, and it's the right way."

KNOXVILLE NEWS-SENTINEL, AUGUST 1, 1971

Controversial Editor—Conservationist Mike Frome Tells It as He Sees It

By Sam Venable Jr.

He's just as likely to show up at ladies' garden club meeting as he is at the gathering of a sportsmen's association.

He's as much at home on a remote Smoky Mountains stream as in a Senate committee hearing in Washington.

He might be in blue jeans "chewin' the fat" around a campfire or in a tuxedo dining with the Secretary of Interior.

But wherever he shows up, one thing is certain: Mike Frome will tell it as he sees it.

Sometimes his words leave listeners bitter . . . almost to the point of rage. Other times they are pleased. But no matter what the outcome, he goes his way, all the while promoting the Frome theory.

Conservation editor of *Field & Stream,* one of the nation's largest outdoor magazines, Frome is without a doubt one of the most controversial men in the American conservation scene today.

Admirers think history will remember him as it has Teddy Roosevelt, Gifford Pinchot and Aldo Leopold. Opponents believe Benedict Arnold would be more fitting.

That's because he's outspoken, frank and to the point. Without regard for position, he'll attack—verbally or with the written word—presidents, congressmen, senators, foresters, biologists, loggers, power companies and anyone else he thinks is tampering unnecessarily with nature's bounties.

The spirited Alexandria, Va., writer-author-lecturer visited East Tennessee last week as part of a week long tour of Southern forests and wilderness areas. Here are some points he made to folks he met along the way:

"We're treating the land as a commodity instead of as a trust and as long as we continue to do so the environment will deteriorate. Our resources are running out, but if we would only regard the land as a trust and pluck its fruits conservatively, we'd be happy, healthy and wealthy forever . . .

"TVA's Tellico project, the Army Engineers' Devil's Jump project, the Tellico–Robbinsville road, and the U.S. Forest Service's forestry practices are just a few of many examples of how the 'great outdoors' is going down the drain. The sad part is that the American public is being brainwashed with misleading gobbledygook about costs, benefits and end results . . .

"Forestry as Gifford Pinchot knew it is not the forestry of today. Now, forestry means only one thing—timber production—even at the expense of wildlife, soil and watershed protection and recreation . . .

"We're not thinking of the future of the resource or of the nation. And what kind of legacy is that to leave for unborn generations?"

NASHVILLE TENNESSEAN, OCTOBER 15, 1972

Conservationist Criticizes TVA

By Keel Hunt

The Tennessee Valley Authority should be leading the environmental protection movement rather than using it as a scapegoat when power rate increases are imminent, an outspoken conservationist said here Friday night.

Michael Frome, conservation editor of *Field & Stream* magazine, said TVA is not genuinely responsive to environmental problems and he criticized what he called the agency's "lust to dam every free-flowing river in the Tennessee Valley."

"TVA began with high hopes and low costs," Frome said. "Now the costs are high and they're going up again with TVA saying the cause is environmental protection. It's a low and cheap trick of a cheap and low agency," he said.

Officials of the agency have said repeatedly that long-range costs of protecting the environment from sulfur dioxide emissions—a by-product of burning coal—and other pollutants could total as much as $800 million over a 20 year period.

"TVA should be in the lead of the environmental movement and not throwing these brickbats at it," said Frome, who is in Nashville to address the Tennessee group of the Sierra Club. . . .

DENVER POST, FEBRUARY 8, 1972
U.S. Leaders Criticized on Environmental View
By Dick Prouty

The power structure in society has lost touch with the people, Mike Frome, nationally known writer and conservationist, said Monday in Denver.

The political and corporate leadership doesn't understand the environmental concern of citizens, Frome said as he assailed the Nixon administration for "talking an environmental game" while still playing against the nation's long-term environmental interest.

Frome spoke at a meeting of the Wilderness Society and the Colorado Wilderness Workshop at Denver Botanic Gardens auditorium.

Asked what people can do for the environment, Frome, a columnist for *Field & Stream* magazine, suggested stopping wasteful, conspicuous consumption, recycling, cutting back on use of material goods, backing closing of factories that claim they can't afford to solve pollution problems, probe the "phony financing" of many government projects and supports to private industry, respect life forms whether obviously useful to man or not. He predicted the "extremism of today will be the way of doing business tomorrow." . . .

"A conservationist isn't trying to solve today's problems," he observed, "but those of 100 years hence, or even further. He's looking ahead to the future to be sure there's a land left to yield the rich resources of the earth to a future generation. He's an optimist in the face of overpowering problems."

BILLINGS (MONT.) GAZETTE, JUNE 17, 1973
Fuel Crisis Phony, Writer Says
By Michael Wenninger

The "energy crisis" is phony and the Nixon Administration is polluted, according to a tough-talking magazine writer who stopped in Billings while looking around Montana last week.

Michael Frome, conservation editor of *Field & Stream,* remarked, "We shouldn't allow the power structure to dictate to us on the so-called energy crisis, for which there will be absolutely no end unless we change the national direction at once and make a long-range commitment to eliminate waste and restrict growth and affluence."

That was one of the more eloquent statements by Frome during an interview. While verbally wiping out everybody in the Administration who has anything to do with the environment, most of his remarks had the tone of this one:

"Reclamation [of strip-mined areas] looks to me like putting the mask on a ghost."

Frome pounds on the government in his articles in *Field & Stream* and he has developed an influential following.

It is possible he was responsible for the defeat last November of Colorado Rep. Wayne Aspinall, a powerful ally of the mining industry in his position as chairman of the House Interior Committee.

Before the election Frome rated all the congressional candidates according to their stands on environmental issues. The published ratings included a special category:

"The man who absolutely must go: Wayne Aspinall."

Frome, who lives in Alexandria, Va., flew to Billings Wednesday on the airliner that brought Interior Undersecretary John C. Whitaker. At the airport, Whitaker told reporters that the Administration's mining reclamation bill is a tough one, the energy crisis threatens us, the Alaska oil pipeline is needed, and the Northern Great Plains Resource Program will help control coal development.

The next day Frome commented on Whitaker, noting he had been on the White House staff and had drafted President Nixon's 1972 energy policy "that started the ball rolling for all energy uses."

"All of his (Whitaker's) actions have been for the benefit of commodity interests and not for the little man of America," Frome said. "The grassroots, not Washington, can handle the fate of these desperate and critical times."

He charged the Northern Great Plains Resource Program [of the Department of the Interior] is disorganized and ineffective.

"I think the best thing we can do with the oil in the North Slope (of Alaska) is to keep it in the ground until we develop an ethic of sound use," Frome said.

He remarked later, "We get the word that there's an energy crisis from the people in the Administration and the big oil companies. "Claims of a shortage of oil are designed to raise prices and to sanction rape of the resource." . . .

The conservationist said a bright spot in his Montana visit was the news that a coalition of environmental organizations had sued the federal government in an attempt to halt coal development until its impact is determined. "That shows the voice of the people can be heard. And considering the power of the vested interests in Montana, it's a real uphill battle. The power structure must be shown there are other values than just coal to be mined in the eastern part of the state."

Frome came to Montana to speak Saturday at a meeting of Trout Unlimited's state council and national directors at West Yellowstone. Last year Trout Unlimited named him "Trout Conservationist of the Year."

STILLWATER (OKLA.) NEWS-PRESS, SEPTEMBER 9, 1973

Field & Stream Editor Defends Fired Biologists

By Fred Miller

Gov. David Hall is reluctant to investigate charges of incompetence and political favoritism in the state wildlife department, says Michael Frome, the outspoken conservation editor of *Field & Stream* magazine.

Frome, speaking on the OSU [Oklahoma State University] campus Friday, praised as "heroes" the three wildlife biologists fired from the department in June for "embarrassing the people in power."

The editor revealed he had spoken with Gov. Hall Friday morning and had urged him to use his position to upgrade the department, which ranks as one of the poorest state wildlife programs in the nation.

Gov. Hall admitted that the agency was in trouble, Frome said, but was unwilling to appoint the suggested committee of environmentalists and wildlife professionals to study the problem. Earlier reports about the controversy suggest that Hall is fearful of offending a powerful group of state senators.

Tom Hines and Gene Stout, two of the ousted wildlife researchers, were in the audience as Frome made his remarks. They and Tom Eubanks, the other man fired, claim a report they filed on deer control showed the incompetence of their superiors.

The wildlife department says they were released primarily as the result of budget cuts.

"Either we are going to have wildlife biologists operating without fear of political favoritism or we might as well give up the whole thing," Frome said. Citing another state controversy, Frome warned Oklahomans that some officials in the Weyerhaeuser Corp. look upon environmentalists as "kamikaze conservationists." Weyerhaeuser, the nation's largest timber concern, recently purchased more than 900,000 acres in Southeastern Oklahoma and 700,000 acres in neighboring Arkansas and are now employed in an extensive program of clear-cutting to eliminate the "inferior" native trees and replace them with a "genetically superior" and faster-growing type of pine. . . .

Foresters are directed to "keep the wood-basket filled and net profits high," Frome said. He added that foresters are often told to exceed safe limits in clear-cutting, and seldom seek help from wildlife, recreation and fisheries professionals before making decisions which affect large forested areas.

The editor fielded several sharp questions from the audience, most of whom were wildlife, forestry and biology students and faculty. "Many forestry schools wouldn't let me in," Frome said later.

Frome indicted the Nixon administration for its poor environmental record. "I think the decisions are being made in the White House, and, unfortunately, the heads of the Forest Service are passing those decisions along." . . .

Following his speech Frome lunched with Hines, Stout and a few faculty members and wildlife professionals. He was scheduled to speak Saturday at a Wildlife Federation meeting at Lake Texoma.

MOUNTAINEER, WAYNESVILLE, NORTH CAROLINA, AUGUST 9, 1974

Noted Conservation Writer Backs Cataloochee Committee

A lawsuit filed 10 days ago by the Committee to Save Cataloochee Valley has apparently stalled progress on the proposed access road through that area, according to Michael Frome, conservation editor of *Field & Stream* magazine and author of the book "Strangers in High Places."

"I've never seen citizens act so admirably as in Western North Carolina," Frome told a press conference at the Maggie Valley Holiday Inn Wednesday. He said that as a result of the suit, National Park Service Director Ronald Walker indicated to him earlier this week that construction of the road would not begin until the suit is settled.

Frome and Toby Cooper, of the National Parks and Conservation Association, came to Haywood County from Washington, D.C., to lead a hike through part of the valley and to speak at a banquet for concerned citizens.

"I was here for the transmountain road fight in 1966," Frome said, "and I'm sorry to have to be here for a rerun. But if we licked them once, we can do it again."

Frome said the National Park Service did not ask for public input on the issue and that the NPS position is nothing more than a justification of prejudgment to build the road.

"When you give a federal agency money to construct something," Frome said, "they're going to find something to construct." . . .

Asked about the safety factor of the present road in the valley, Frome said he did not think the gravel road is a safety hazard. "I think those dirt roads are beautiful," he said. "That's what they should have in national parks if they must have roads."

(SAME NEWSPAPER, SAME DAY)

Time Out

By Mike Jones, Sports Editor

Hiking through the Cataloochee Valley with Michael Frome, conservation editor of *Field & Stream* magazine, brought an interesting discovery to light.

After hiking for some time, someone mentioned that many "gallons" of corn were once produced in the valley, and it later came to my attention that Frome is an apparent connoisseur of the liquid variety, actually preferring it over the newfangled, store-bought stuff.

When asked what he mixes his corn squeezin's with, an expression of pure horror knitted his brow, and he commenced to explain that any mixing would absolutely ruin moonshine, as it is made of corn—the most desirable ingredient in a whiskey.

Drinking it on the rocks, though, is okay, he admitted.

Frome said he only uses the firewater for special occasions. Do you suppose they really do have barn-raisings and hoe-downs in Washington? Maybe since Nixon resigned [one day earlier]. . . .

SALT LAKE TRIBUNE, SEPTEMBER 22, 1975
The Monday Environmentalist

Dauntless Conservation Writer Pays Dearly for Candor

By Ernest H. Linford

Because they regarded it too critical of the U.S. Forest Service the directors of the American Forestry Association late in 1971 voted to drop Michael Frome's column from the magazine *American Forests.*

Frome, author of several authoritative books and widely regarded as an expert on environmental matters, had written this main feature for *American Forests* for years. But timber industry–dominated directors complained that "Mike gets us into trouble," and he was banished.

Undaunted, Frome continued naming names and calling the shots as he saw them, mainly in his "Conservation" column in the magazine *Field & Stream,* devoted to hunting-fishing.

In 1974, *Field & Stream* dropped Frome's column and replaced it with a bland substitute that avoids grips with real conservation issues. As a result, Columbia Broadcasting System, which owns *F&S* [and later sold it to the *Los Angeles Times*], was the target of angry letters, phone calls and demonstrations. CBS denied any part in Frome's firing, but it had passed the word down "not to name names."

Frome's freelance writing appears in other magazines but the loss of two outlets with national circulation must have hurt. No critic has been able to pinpoint factual errors in his articles, but he is controversial. He angers the businessmen and politicians he lists as enemies of conservation.

The other day a student [at the University of Wyoming, where Linford headed the Journalism Department], returning to school by way of Chicago, brought back a copy of a recent Frome article in the *Chicago Tribune,* of all publications. It was about "political encroachments" of the national parks. And it demonstrated that the "old master" has not lost his touch. And he still names names!

An example is the case of the horrendous 307-foot tower, a commercial enterprise which now dominates Gettysburg National Military Park in Pennsylvania. It was made possible through political influence. The family of the tourist enterprise promoter, Thomas R. Ottenstein, contributed heavily to the Nixon campaign. And construction financing was through I. H. Hammerman, crony of former Vice President Agnew.

The tower stands on private land adjacent to the park but is serviced by a road through the park. Former Interior Secretary Morton at first branded the proposal for it "the most damaging single intrusion ever visited upon a comparable site of American history."

Three weeks later Morton quietly signed over the right-of-way for the commercial outfit, even though the state of Pennsylvania, historical societies and conservationists protested the action.

Another scandalous condition involves park concessioners who furnish lodging and other services in national parks under federal contract, says Frome.

A focal point is Yosemite National Park in California. Two years ago the concession there was acquired by Music Corporation of America, a Hollywood conglomerate with political clout with plans to transform Yosemite Valley into a money-making resort. MCA officials were in on the master plan at Yosemite even while interested public groups were kept in the dark.

Jay Stein, MCA head, undertook to bypass even the park superintendent and regional director through connections in Washington. MCA hired William Ruckelshaus, former head of the Environmental Protection Agency, to represent it in Washington. "Little wonder," says Frome, "that park officials are always cautious, sometimes even timid, about their responsibility. Concessioners exercise more power in the parks than park superintendents."

He cites other examples of special interest attempts to seize irreplaceable public treasures, such as the Disney-in-the-Sierra commercial plot to take over Mineral King, Calif., a fragile alpine valley virtually encircled by Sequoia National Park, strip mining in the heart of Death Valley National Monument in California, and efforts of utilities to run a corridor of power lines through Glacier National Park to join up with giant strip mining and oil and gas enterprises nearby.

A giant electric utility complex under way in the Southwest, including the Kaiparowits project in Utah, threatens to smoke up and otherwise degrade what Frome says is one-fifth of the whole National

Park System, as well as spectacular Monument Valley, Shiprock and the Hopi and Navajo Indian reservations.

Can you hear someone complaining, "Mike gets us into trouble?"

MISSOULIAN, MISSOULA, MONTANA, JANUARY 7, 1983
Frome Continues the Fight
By Dale Burk

A decade ago writer Michael Frome of Alexandria, Va., was fired as a columnist on conservation issues from two of the nation's better known publications—*Field & Stream* magazine and *American Forests,* the journal of the American Forestry Association.

Editors at both publications took exception to Frome's caustic and, it turns out, incisive and accurate assessment of mismanagement on the national forest system.

Frome was in Montana recently to carry on his crusade for better natural resource management. And he was neither repentant nor angry that those two publications had succumbed to industry pressure and let him go. It was just something that goes with the turf when you take on the role of critic and commentator, whatever the medium and whatever the subject.

Part of Frome's subject, then and now, was the Bitterroot National Forest, wilderness, wildlife, and humankind's attitude toward them. He gave the keynote speech at the recent Montana Wilderness Association convention and told Montana wilderness enthusiasts that the best they can hope for is to save some of America's still wild land for posterity.

"We're not looking at how much more wilderness we will have, but how much less of it," Frome said.

His major theme, as it has been for years, was individual responsibility—in government, industry, university, and each individual's private life. Frome the speaker and Frome the writer key on the same basic premise: the American political process requires participation by those who care about things. And he chided those who think that involvement only to the point of passing a law is sufficient.

"Wild places don't save themselves and laws don't save them either," Frome said. "Regulations have their place, but it's only by a built-in sense of individual responsibility that anything is saved."

Nowadays Frome writes a regular column, titled "Crusade for Wildlife," for *Defenders of Wildlife* magazine, keeps a heavy speaking schedule, and is at work on a couple of books—including one tentatively titled *The Truth About Our National Parks* [which became *Regreening the National Parks*]. He also will see an updated version of his classic *The Forest Service* come out in 1983.

Frome also will be much more involved in resource issues of the Northern Rocky Mountains. He is spending this school year at the University of Idaho as visiting associate professor of communication and wildland recreation management.

It's a teaching position he cherishes, coming as it does on the heels of a year spent at the Pinchot Institute for Conservation Studies at Milford, Pa. He was author-in-residence there for a year and spent the time speaking and writing his book about the Forest Service.

Now he's convinced that the national debate over wilderness will teach much about the need for resource conservation.

"We modern Americans have yet to fully realize the notion of a non-materialistic approach to the earth," Frome said.

He sees the wilderness debate that has come to sharp focus in the Northwest and the Bob Marshall Wilderness in particular as "a fitting place to review and reassess wilderness as a component of our life."

Frome is helping to both broaden and sharpen the questions central to that debate. Issues involving a particular wilderness or policy are important, but are only part of the equation. He is correct in suggesting that we should be asking, now, if we really understand just how much wild places do to enrich the civilizations of which they are part. And he is correct to have stayed the course on his crusade for things wild.

Mike Frome is a perfect example of the axiom that to stifle debate on controversial issues is ultimately self-defeating.

Albuquerque Journal, September 22, 1985

Wilderness Conflicts: Rangers Go to School

By Nolan Hester

Henry David Thoreau once wrote, "In wildness is the preservation of the world." Nice idea, but who preserves the wilderness itself?

The common notion is that wilderness just sits there—undisturbed, primeval. In fact, hikers and horses thread its trails, fires sweep or sputter in the woods, cattle graze the meadows.

Across the Southwest, balancing demands on the land often falls to the U.S. Forest Service, the region's biggest wilderness landlord. To help rangers keep the wilderness wild, the Forest Service sends them to school.

The school's classroom—Northern New Mexico's Pecos Wilderness, 223,000 acres of alpine peaks plunging to sapphire lakes, mountain meadows and rippling trout streams.

The students—33 Forest Service employees from the agency's far-flung Southwestern empire. There are also rangers from Idaho and Wyoming and a handful of people from the National Park Service and Bureau of Land Management. To poke holes in agency theories, the Forest Service also invited conservation writer Michael Frome.

The curriculum is straightforward. If rangers are to understand the wilderness and its users, they must head for the hills and take to the woods. For a week, they leave desks and uniforms behind, carry backpacks, get wet.

As the class unloads gear at the Jack's Creek trailhead, old acquaintances are renewed and introductions evoke a New Mexico–Arizona map full of memories. "That's some great country you've got over there," they murmur to each other knowingly.

Just how to keep it great is an open question.

Should certain areas—the tiny mountain lakes that draw 95 percent of the visitors but cover 5 percent of the wilderness—be sacrificed to heavy use? Should campers be forced to spread out, perhaps taking trash even deeper into the woods?

The Forest Service leaves each district to come up with its own solutions. Director Floyd Thompson says the school's aim is to get people to trade ideas. Relishing his role as provocateur, Frome hopes a few students will agree with him that the Forest Service may be as much the problem as the solution to wilderness conflicts.

Heading up the trail, the image of wood-wise rangers is quickly shattered. For many, this is their first real hike in years. They huff and puff even though an outfitter has already packed in their food.

Rangers don't backpack, they ride horses. "Distances like these are why God made horses," says Bill Russell as he considers the four miles to base camp.

Still, the format propels them along—hike and talk, hike some more and talk again.

Everything from trail construction to firefighting, air pollution and campsite abuse gets debated during a series of trailside stops led by agency specialists. The rangers debate the exact pressure a cow's foot exerts on a meadow—such details are the currency of ranger conversations.

Yet they cannot name shrubby cinquefoil, a common Southwestern plant. And there is little talk of the surrounding country's sheer beauty.

For Frome, the incident spotlights the agency's blind spot—all head, little heart. He complains that the Forest Service sees wilderness as yet another commodity to be used. Like Thoreau, Frome sees wilderness as far more than that.

"Wilderness is the heart and soul of America. All our art, literature, poetry derives from the natural world," he says one night by the fire. Wilderness makes bureaucrats people." . . .

JUNEAU (ALASKA) EMPIRE, JUNE 23, 1985

Professor Looks at Negative Effects of Tourism on Southeast

By Chuck Kleeschulter

Cruise ships and the thousands of tourists they unleash on Southeast may not be the clean, non-polluting industry that tour companies like to claim, according to a professor and environmental journalist.

In fact, it may be time to limit and possibly reduce the number of cruise ships stopping in Southeast ports and firmly cap the numbers entering Glacier Bay National Park and Preserve, said Michael Frome, the former editor of *Field & Stream* magazine.

"Cruise ships shape the lifestyle of a community. If you have a handful of tourists off the boats they are visitors. But if you have a thousand they become numbers and that causes the attitudes of local residents to change. They begin to affect the lifestyles and environment of the communities they enter," said Frome, a professor at Huxley College of Environmental Studies at Western Washington University in Bellingham. . . .

Frome said cruise ships tend to increase air pollution, water pollution and marine pollution if crew members "take the easy way out" and dump garbage overboard. In their current numbers the ships also are visually polluting the environment, he said.

"I know I wouldn't want to be hiking in the backcountry at Glacier Bay and look down and see several large ships in the bay. Visitation to the bay has exploded and it's time to consider what that is doing to a national park whose primary purpose is for the preservation of the glacial environment," said Frome.

BELLINGHAM (WASH.) HERALD, JULY 11, 1989

Environmentally, Bellingham Is Headed down the Tube, Says a WWU Professor

By Francine Strickwerda

Bellingham may be a wonderful place to live, a nationally known environmental activist said Monday, but the city is in a condition of galloping environmental decay.

"Bellingham and Western Washington are fast losing the beauty that has been cherished forever," said Michael Frome in a speech to the Bellingham Rotary Club.

Frome, who is also an author and professor at Western Washington University, said the city and What County governments operate on the principle of "anything for commercial industry." . . .

CHATTANOOGA NEWS-FREE PRESS, JULY 29, 1989

Green Party Advocate Rates Bush Low on Environment

By Steven Epley

President Bush has "fallen down" when it comes to protecting the environment, and the time is ripe for a populist political party like a European Green Party to arise and make government more sensitive to ecological concerns, a nationally renowned conservationist and author said here Friday.

Michael Frome gave the president "good grades" for "reversing the course of the Reagan administration," which he called "undoubtedly the worst administration in terms of the environment in the 20th century, if not in the history of the republic."

"George Bush is no Teddy Roosevelt," the former conservation editor of *Field & Stream* told the Chattanooga Civitan Club. Mr. Frome said the president reacted "passively" to the Exxon oil spill in Valdez, issued a "weak proposal" to clean up the air and water and has appointed "unqualified hangers-on to important environmental positions as political payoffs." . . .

"People don't trust the government any more and it's about time that it be responsive to the critical needs of our time and the needs of people," said Mr. Frome, who also spoke Friday night at an Earthworks-sponsored event at the Hubert O. Fry Center in the Tennessee Riverpark. . . .

In a news conference between his speeches, Mr. Frome said, "We need to organize a Green Party in this country because I don't see Republicans or Democrats giving adequate response to environmental problems."

Many European nations have small Green parties, several of which are growing in numbers and importance as they demand greater governmental action to protect the environment.

Reminded that third parties in this country have traditionally had a hard time attracting voters and favorable publicity, Mr. Frome said, "Yesterday's way of doing business doesn't necessarily hold for tomorrow. Just because there's not been a successful third party doesn't mean there can't be a successful Green Party.

"At every lecture I deliver, people say to me, 'What can I do to help?' said Mr. Frome. "The reason they come to me is that the government doesn't want them to help."

Pointing to Mr. Fry, longtime local conservationist seated in the audience, Mr. Frome said, "Hubert has told me about the crud floating down the river and ending in the riverport. It's being dumped upstream by thoughtless people, not mean people."

Champion Paper Mill in Western North Carolina, which has been asked by Tennessee Gov. Ned McWherter to stop polluting the Pigeon River with its emissions, should "clean up and not try to out-PR the governor," said Mr. Frome, using the initials for public relations.

In Canton, N.C., near the Champion plant, there used to be a saying that the smell of pollution was "the smell of green," meaning money said Mr. Frome. "There must be a better way. It's too high a price to pay, and the jobs aren't there anyway because they're substituting machines for men and women."

Mr. Frome said Tennessee is doing "very poorly" in its fight against pollution. "Tennessee is a major source and victim of acid rain. Air pollution here is more severe here than elsewhere, and the landscape is more scarred by billboards than in other states."

ASHEVILLE CITIZEN-TIMES, JUNE 22, 1975

Conservation a Success in Mountain Area— 'There Is Hope' Has Been More than a Slogan

By Jim Dean

Raleigh—I know that lots of people believe the world is going to Hades in a handbasket, but I suspect that's nothing new. The human animal has always been an instinctively pessimistic critter. Cavemen probably thought the world was going to Hades in a basket too.

Anyway, I wonder if I might be forgiven for injecting a ray of hope into these otherwise dismal times.

This past week, I listened to a lot of knowledgeable folks discussing such matters as land management, National Park expansion, wilderness protection, nuclear power, energy conservation and all sorts of other things.

But out of all of this, the one line that rings in my mind like a bell is something conservation writer Michael Frome said.

"There is hope," said Frome. "There is always hope."

Then he proceeded to point out some of the changes that have not yet occurred in Western North Carolina. He mentioned the notorious transmountain road from Bryson City to Townsend, Tenn. A few years ago this road seemed vital to a lot of people and it would surely have ruined the wilderness aspects of much of the nation's most popular national park. Now such a road seems highly unlikely.

Then Frome mentioned the controversial Tellico Plains-Robbinsville road that would separate the Joyce Kilmer Memorial Forest from the wild reaches of Slickrock Creek drainage. Creeks, virgin timber and many wildlife considerations were in jeopardy, but it now appears that this area may be protected against any road by the Wilderness Act.

Frome then mentioned the 14 dams proposed on the French Broad watershed by the Tennessee Valley Authority before strong and united public opposition finally forced TVA to retract its plans.

As Frome pointed out, in each case it seemed hopeless to resist these projects because they were planned by powerful forces, and their completion seemed inevitable. But enough people cared about them to get in and fight. As Frome said, "there is always hope."

Now you may not agree that all—or any—of these things were worth saving, but that's not really the point as I see it. The point is not how you feel about certain environmental matters (nuclear power, controversial roads and dams, oil exploration, ozone, flood plains and many others).

The significant thing is that these matters are now being openly considered and questioned in terms of their effect on our environment!

That is what the environmental fight has been all about for the past decade—to gain the objective of openly planning and evaluation BEFORE they are carried out instead of waiting too late to find out what the results might be.

Why Downsizing instead of Upsizing?

Celebrating the seventy-fifth anniversary of
the National Parks and Conservation Association,
San Francisco, May 18, 1994.

My friend Paul Pritchard proved to be a talented go-getter as president of the National Parks and Conservation Association. He built the organization, raised funds to expand the staff, and ran good conferences that I was privy to attend. The 1994 celebration in San Francisco was one of them. Besides the program at which I spoke, it included a banquet attended by environmental leaders and luminaries. In the speech that follows, I said that when I come to California I think of David Brower, Edgar and Peggy Wayburn, Bill Lane, and Martin Litton, "Bill's favorite curmudgeon and mine." I believe all of them were at the banquet. I remember for certain that Martin was there because during the reception he pulled out a set of photos to show me his appendectomy, with his stomach open in living color, until his wife said, "For heaven's sake, put that away, Martin!"

In the preface I mentioned L. W. (Bill) Lane, the retired publisher of *Sunset Magazine,* for whom Martin had worked as travel editor. Litton in those years was already an activist in league with Brower. When he turned in an article about the controversy over the Redwoods, Lane rejected it as too radical for *Sunset,* so Litton got mad and quit. His assistant succeeded him, went over the Redwood story and showed it to Lane, who was pleased. "Now that's the way it should be," he said, although it was exactly as Martin had written it.

Martin Litton *is* my favorite curmudgeon. When I was at the University of Idaho, I invited him to come speak to my journalism class. He flew up from the Bay area in his own plane at his own expense, then set the students straight on the question of objectivity based on his early experience writing for the *Los Angeles Times.* "I always gave both sides," he told the class, "and then I said which side was right and which side was wrong."

■　■　■　■

For many years national parks have been an important part of my life, continually inspiring, rewarding, and challenging, more now than ever. I view national parks as sanctuaries, sacred space, ideal places for escaping cities to look at stars—for touching stars and being touched and empowered by them. National parks open the heart to inner feeling and emotion; they enable me to appreciate the sanctity of life, of all life, and to manifest humility and love as a child of God's own universe.

John Muir wrote that wildness is a necessity, that mountain parks and reservations are useful not only as fountains of timber and irrigating rivers, but as fountains of life. Theodore Roosevelt in 1903, after camping with Muir among the ancient sequoias of Yosemite, listening to the hermit thrush and the waterfalls tumbling down sheer cliffs, wrote that "it was like lying in a great solemn cathedral far vaster and more beautiful than any built by the hand of man."

National parks evoke such response from the human spirit. They are the landscape meant for poets and painters and all types of creative persons. "Here," said Ansel Adams, "are worlds of experience beyond the world of aggressive man, beyond history, beyond science."

Because I love the national parks and have worked long on their behalf, I feel deep concern for their present and fear for their future. National parks constitute a gallery of American treasures—an endowment of riches that makes the United States the envy of the world—but they are not well protected. They are at risk. Our national parks, monuments, and historic shrines are in trouble. They cry for understanding, help, support, and advocacy from citizens who care, precisely the kind of citizens the National Parks and Conservation Association has brought together in San Francisco. I feel honored to be here with you, to have this opportunity to review traditional values as the foundation of the new agenda I believe critical for our national parks.

"Each one of these national parks in America is the result of some great man's thought of service to his fellow citizens. These parks did not just happen; they came about because earnest men and women became violently excited at the possibility of these great assets passing from public control." So declared J. Horace McFarland, as president of the American Civic Association, in testifying before a congressional committee in 1916. Years later, another Horace, Horace Albright, who devoted his long, illustrious life to the parks cause, added the hope that more and more people as individuals would share in preserving our heritage: "Many, through voluntary association, can give material support to conservation projects together that they could not give separately. Legislators intelligently concerned with conservation deserve and need wider support from more citizens who will take the trouble to inform themselves of new needs and weak spots in our conservation program."

I myself faced a revealing "weak spot" last month in Colorado, where I spoke on a panel together with regional directors of the four federal land-management agencies—the Bureau of Land Management, Fish and Wildlife Service, Forest Service, and National Park Service. Each one of them in turn talked of "downsizing," of scaling back activities as a consequence of budget reductions. I found that distressing. Consider there is no downsizing in building prisons, or in programs dealing with crime, drugs, and violence, or in searching for a cure for AIDS. National defense may be pared around the edges, but the budget is still in the neighborhood of $300 billion. Those plainly are identified priorities of our time. But if open space and recreation and biological diversity and environmental health and all those good things are valuable and valid in the public interest, why are these federal land agencies downsizing instead of upsizing?

The regional directors provided at least part of the answer in the manner of their presentations. They talked in rhetoric and jargon, of structure, rather than of substance, or challenge or hard issues, promising to do better but without any system of genuinely informing and involving the public as partners in protecting the public estate.

I found further evidence of "weak spots" in an article in the *Washington Post* of May 10, 1994, barely one week ago. The article began as follows: "Stung by criticism of a land swap gone sour at Gettysburg National Military Park, the director of the National Park Service yesterday promised to freeze future property deals while writing stringent new guidelines to protect other federal sites."

Why in the world do these promises almost always come after the fact? I don't mean to hold the director, or any individual, personally responsible, but the last paragraph of the article shows something seriously systemically wrong here. It quotes the chief historian of the Park Service as saying he first learned of the damage at Gettysburg in a 1991 phone call from a congressional staff member, but he didn't object because his previous protestations over intrusions at Vicksburg and Petersburg battlefields had been ignored by decision makers within his agency.

How sad. It takes commitment, conviction, deep-felt concern, and courage at all levels, top to bottom, to run the national parks and to relate to caring citizens. Stephen T. Mather, the first director, built a field force as a model of honorable and ethical federal employment. He inspired the National Park Service "mystique," a spirit of mission, a willingness to stand tough against what he called "desecration of the people's playground for the benefit of a few individuals or corporations."

Horace Albright, as superintendent of Yellowstone in 1920 defined criteria for the position of ranger. The ranger, he wrote, must be tactful, diplomatic, and courteous, with outdoors experience in riding, camping, woodcraft, and fighting fires. The pay was only one hundred dollars per month, and the ranger must pay for his own meals, buy his own uniform, and "bring his own bed." Many brought their own horses as well. Albright made it clear there would be no soft and easy jobs in the National Park Service. "The ranger's job is especially hard," he advised applicants. "Apply if you are qualified. Otherwise, please plan to visit the Yellowstone National Park as a tourist."

National Park Service personnel need to learn more history and how to apply it to their lives and their careers. They need to learn about the citizen heroines and heroes who made the parks happen and have worked diligently to protect them. National parks bring out the unselfish, ethical best in people and these citizens are the proof of it.

I will cite a few as tribute to the many, starting with George Bird Grinnell, most closely associated with Glacier National Park. His circle of friends ranged from rough frontiersmen and Indians of the teepee to scholars, statesmen, and presidents. When Calvin Coolidge in 1925 presented him with the Roosevelt Memorial Medal, the president told Grinnell, "Few men have done as much as you, and none has done more, to preserve vast areas of picturesque wilderness for the eyes of posterity in the simple majesty in which you and your fellow pioneers first beheld them."

Another, William Gladstone Steel, from the first time he saw Crater Lake, in 1885, worked tirelessly, building support and scientific data and influencing Oregon's congressional delegation until the national park was established in 1902. Later he said, "Aside from the United States government itself, every penny that was ever spent in the creation of Crater Lake National Park came out of my pocket and, besides that, it required many years of hard labor that was freely given. . . . All the money I have is in the park, and if I had more it would go there too." In contrast to other great American lakes, like Tahoe, that fall in the category of paradise lost, Crater Lake, thanks to Steel, retains its natural character, a vestige of the original America.

Harry F. Byrd was governor of Virginia in the 1920s, when Shenandoah National Park was authorized. Every acre was in private ownership and Byrd played a key role in working with citizen advocates and the legislature to acquire the land for presentation to the federal government. Years later, as a U.S. senator, he would write, "In the tragedies and other strain of our modern world generations to come will receive peace of mind and new hopes in lifting their eyes to the peaks and canyons of the Shenandoah National Park and those who made possible its establishment can justly feel that their labors were not in vain."

Marjory Stoneman Douglas awakened fellow Floridians and other Americans to the wonder of the Everglades with her book, *River of Grass*, and has never quit working to protect the park. In 1986, when the National Parks and Conservation Association presented me with the first Marjory Stoneman Douglas Award—the first, that is, after she herself received it the year before—Marjory came to the program in Washington and stole the show. But she was entitled, at age ninety-six still going strong and working for the future.

Gilbert Stucker in 1953 made his first visit to the most famous dinosaur repository in the world, Dinosaur National Monument, in northeastern Utah and western Colorado. He was a professional field paleontologist but came primarily as a citizen preservationist concerned with proposals to construct dams across the Green and Yampa Rivers within the national monument. Stucker proposed using the quarry as a positive project, a platform from which to generate public interest and concern. He convinced the National Park Service to build the unique visitor center displaying and interpreting tremendous bones of animals 140 million years old etched out in relief, right in the rock where they died before the rock solidified. Then he accepted temporary appointment as ranger-naturalist, telling visitors about ocean beds back in the canyons with millions upon millions of remains of sea life—shells and corals, coral reefs, all sorts of remains of prehistoric life. He was admonished not to discuss the pending dams, but he felt that he must and did. Thanks to many individuals and organizations, Dinosaur was saved. Stucker continued his conservation work, joining efforts to add choice areas to the national park system, including Fossil Butte National Monument, in Wyoming, preserving fish forty or fifty million years old in laminated clays and silts. Stucker became active in the National Parks and Conservation Association and served a term as chairman of its board.

Citizen organizations and individuals have provided the enthusiasm and energy to establish and then to guard the integrity of national parks. When I visit Dinosaur, the Grand Canyon, the Redwoods, the North Cascades, I think of David Brower, with gratitude for his leadership and inspiration. When I come to Golden Gate, the Redwoods, and the Alaska parks, I feel indebted to Edgar and Peggy Wayburn. I think of Bill Lane's important contributions through the years at *Sunset,* and of Bill's favorite curmudgeon and mine, Martin Litton. In the Southeast, Liane and Bill Russell not only championed the Great Smokies, but were instrumental in establishing the Obed Wild and Scenic River and Big South Fork National Recreation Area.

National parks when they were new yielded discovery, adventure, and challenge. They should always do that. And they can, if we set our minds and hearts to it. As the rest of the country becomes crowded, developed, and supercivilized, national parks should be held apart, safeguarded to represent another side of America, free of technology, free of automobiles, snowmobiles, and flightseeing, free of commerce and crowds, free of instant gratification, a pioneer, self-reliant side of America.

In the unending struggle to obtain and sustain a quality environment of life, national parks contribute by their existence as model ecosystems. In a world where nothing remains static, the challenge is to raise and then sustain the quality of the parks themselves. Turning over the National Park System in better condition than we found it may be the most important bequest of this generation to the future.

National parks have become a powerful social ideal throughout the world. Virtually every country needs and wants them. Parks certainly make better calling cards than bombs do and contribute more to peace. The rest of the world looks to the United States, where park systems are most advanced, and the United States must not betray that trust.

But how can we do this in face of the downsizing I mentioned earlier? For one thing, I urge more emphasis on natural sanctuaries as human sanctuaries, manifestations of human dignity and mutual respect, reminders that we are all connected, as brothers and sisters of common origin, common destiny, children of the same Creator. I can't escape the truth that the poor, especially the nonwhite, are disenfranchised from the bounties of our time. One third of America's black population lives below the official poverty line. More than 60 percent of all black births are to single women, and almost 50 percent of black children are being raised in poverty. Economic devastation results in social disintegration. Inner cities and inner-city schools are so dangerous that most law-abiding people and the children who want to learn, and teachers who want to teach, lead lives of terror.

A system that places a low priority on human values and natural values needs to reorder its priorities. Americans should face the twenty-first century with new emphasis on human care and concern. Those children need and deserve hope, a reason to believe, to study for a future with meaning. National parks and other types of parks should serve their interest. I recall the experience of three inner-city black teenagers in Philadelphia hired through the Urban Coalition for a summer work program at Bartram's Garden, the oldest botanical garden in America. "The other kids couldn't understand why we were doing it," one said. "They were sitting around doing nothing, bored all day, while we were out learning, doing something worthwhile." And from another: "When I walk down the street I can point to certain trees and give their names. I couldn't do that before. I have a better understanding of nature and how it works. I relate to Mother Nature, and she relates to me."

Something new must be done by a new group or groups of people. The priority item on the agenda is for those who hope to heal the earth to join with those who hope to heal the souls of our fellows to bring something new to bear. "New opinions are always suspected, and usually opposed," wrote John Locke more than three centuries ago, "without any other reason but because they are not already common." Such is the way of institutions, but not of individuals. Even in our age of stereotypes, true success comes only from within. When we look at the revolutionary task of reordering priorities, and the sheer power of entrenched, interlocking institutions, the challenge may seem utterly impossible. Yet individuals working together, sometimes even alone, have worked miracles. That is the history of our national parks and the history of our country.

In setting the national parks agenda for tomorrow, miracles large and small are within reach. Theodore Roosevelt when he spoke at Stanford University in 1903 said there is nothing more practical than the preservation of beauty, than preservation that appeals to the higher emotions of mankind. Yea verily, those who safeguard a national park as a sacred cathedral of harmony and hope will be blessed.

For Students with
Daring Dreams

Commencement exercises,
Alger Learning Center/Independence High School,
Alger, Washington, June 17, 1995.

I spoke at a few commencement exercises, not many but enough to learn how *not* to do it.

In 1988, I was invited to give the commencement address at Huxley College, where I had come to teach the year before. I prepared carefully and thought I had something special to say. I started with a few jokes and stories. They were well received. The audience was in a good mood and I probably should have quit while I was ahead. But I plowed on with a serious message. Halfway through, I realized the audience was no longer with me; I was strung out on a line all by myself. I struggled through, cutting wherever I could.

I learned that graduation is feel-good time. Class is out. A few serious words are in order, but nobody wants to remember the overkill. Keep it a joyous day.

■　■　■

This is a joyous day, the celebration of a landmark in life, a reminder that the most precious gift of all is life itself, thus challenging us as individuals to use life fully, to recognize and appreciate that the great use of a life is to create something that outlasts it.

It is often said that today's students cannot spell, don't know history or geography, and are driven by the desire for self-gratification. That isn't the way I see it in students who have come to my classes over the years, or in young people I have met elsewhere. I don't have to tell you that students are real people, struggling to meet everyday challenges, in classrooms, libraries, and in their lives outside of school, striving to define goals and individual identity. My students care about the world they inherit and the world they will leave. They dream lovely daring dreams, and live them out, too.

It isn't easy in modern, assembly-line American education, where objectives generally are to serve the needs of employers and institutions, not necessarily those of the student. The emphasis in education is on facts and saleable skills, with secondary attention at best to looking within or developing the whole person. Education for the most part is about careers, jobs, and making it in a structured society, but my priority is on cultivating the inner spirit as an essential ingredient in the development of professional skills. I try to open the hearts of students to look inside themselves, believing that individuals can and do make the difference and bring positive change, unlike institutions that breed conformity and institutionalized professions that demand colorless impersonality and presumed objectivity.

Through good fortune in 1987 I was invited to speak at the commencement exercise at Sterling College, an alternative two-year college tucked away in Craftsbury Common, an old New England village in northern Vermont. An anachronism in an age of academic bigness, Sterling encourages students to challenge and analyze prevailing assumptions. In the course of daily field chores, students get cold, wet, and weary, but they acquire conservation skills and appreciation for the life of farmer and logger.

At the commencement exercise, I heard two student speakers open their insides—that was a treat.

"Two years ago I would never have spoken in front of this many people," Lexie Rothstein confessed. "I wouldn't even have introduced myself. I never spoke out unless I had to. I was comfortable keeping all my thoughts to myself. Sterling made me uncomfortable."

The other student, Michael J. Kelly, talked about the Bounder experience, a program of physical challenges. He said it was like digging down deep into himself and finding something in his heart, mind, and body that he did not know was there. It was working as part of a team, asking for help and helping others, solving problems, and overcoming obstacles larger than he could ever imagine. It was, he said, "teamwork, trust, learning about yourself and others, and love."

When the graduates came to the platform for their diplomas, thirty or forty of them, each one in turn hugged the college president, Steve Wright, and many of them cried. I haven't seen that elsewhere. At most commencements graduates accept their diplomas and politely shake the dean's hand, many for the first time.

"Teamwork, trust, learning about yourself and others, and love." I can't think of a better way of defining goals in teaching or learning, especially about the natural environment that is my particular concern. Of all forces in human life, emotion is the most powerful, and of all emotions love is the most powerful. Love is the measure. Students can and often do acknowledge love openly within their peer group, but scarcely ever in the classroom or in dealing with their instructors, or in the professions which they prepare to enter.

This isn't right. I'm for education that illuminates the human condition and defines it in subjective values, essential to personal transformation as the key to transformation of global society. The "subjective" embodies power, the wisdom of knowing the essence of life and caring enough to fight, no longer allowing technical experts to define terms and methods for dealing with critical issues of survival.

Emerson taught that the one thing of value in the world, above all else, is the active soul, and that each person contains within him or her the active soul, although almost always "obstructed, and as yet unborn." I see education as an enriching and elevating pathway that frees the active soul, that engenders love of inner self, that explores the unity and wholeness of creation, the immorality of abusing and exploiting nature for immediate gain, and the benefits of intimate, personal communion with God's earth and sky. The active soul helps to feel the power in one's life.

I don't want to knock the system—that is the easiest thing to do. I feel privileged to be part of it. That I have been free to teach as I do proves that the system works. I tell that to students, too. There is plenty of room for criticism, but far more room to be positive and to give hope. Once following a lecture to an assembly of high school students, one of them in the audience said he had no quarrel with my theme, "But," he asked, "don't you think you sound idealistic?" The best I could say in response was, "I certainly hope so."

The most important legacy our generation can leave is not a world of bigotry, homelessness, hunger and war, but a viewpoint of human distress, concern, and love. It all begins within the individual, inside oneself, with one's own ecosystem, finding the unity of body, mind, and spirit, and reaching out to others to do the same. I believe you graduates of Independence High School have benefited from its alternative approach and have built for yourselves a sound foundation of self-responsibility and self-esteem.

Now I hope you will go on from here to chart your own course through life, defining your own destiny. I hope the experience of the Alger Learning Center/Independence High will encourage you to choose priorities of service and sharing. Believe me, there is no limit to the distance you can travel. "After a few thousand words from her, the world took a new turn," an editor wrote in tribute to Rachel Carson in tribute to her classic work, *Silent Spring*. One lone woman, committed to her research and to clear, courageous expression, made a difference in world thinking.

You will have your chance to make a difference. Remember that the journey toward fulfillment begins inside the whole person. Make the most of it through teamwork, trust, learning, and love.

Parks Are Better
without the Pork

At the annual meeting of the Yosemite Association,
Yosemite National Park, September 16, 1995.

The meeting of the Yosemite Association was like a very large group campout. It was held at Tuolumne Meadows, an assembly of tent cabins at a glorious mountain setting on the east side of Yosemite National Park. The facilities were simple, down to community shower and toilets. It reminded me of how most facilities once were in the national parks and how they ought to be, simple and inexpensive.

The Yosemite Association is one of various groups committed to education about the parks and service to the parks and to park visitors, and composed almost entirely of volunteers who appreciate the outdoors, their very own in particular—in this case Yosemite.

I visited Yosemite for the first time while working for the American Automobile Association. Being in that Sierra wonderland helped shape my idealized image of national parks. I was thrilled at the opportunity to write about them. It seemed to me that hardly any journalists, whether writing about travel or conservation, paid much attention to national parks, so the field was open and I became an expert.

I returned to Yosemite a number of times over the years. In 1968 I came as member of a committee appointed by the Secretary of the Interior to review souvenirs sold in national parks. I had been writing critically of the awful machine-made trinkets purveyed by park concessionaires, trying to make the point that anything sold in a park ought to be educational and park-related, so I was included on the committee. We went around the country, looking at souvenir shops, interviewing park officials and concessionaires, and working to develop our own rational view. In Yellowstone, one member of the committee, Charles Eames, became aggressive about the subject. He was an ally of mine and talked tougher than I did. Eames was a designer of note, best identified for the "Eames chair." "Tell me," he asked the president of the Yellowstone Park Company, "why really do you sell such trash? Would you have any of it in your own home?" The concessionaire smiled benignly. "Not really," he replied, "but, you know, sometimes you do things you don't believe in." Eames snorted and retorted, "I cannot imagine Ansel Adams ever taking a picture *he* didn't believe in!" In due course the committee held its last meeting at the Ahwahnee Hotel in Yosemite. We turned in our report calling for higher standards governing park souvenirs, but nothing really changed.

I came again in 1984 to meet with park interpreters in advance of the summer season. Most interpreters were seasonal employees who normally worked as high school teachers or college professors. In an earlier speech ("Replacing Old Illusions with New Realities"), I cited the case of Alfred Runte, who was rebuked and assigned to "rehabilitation training" for his outspoken interpretive lectures in Yosemite. Runte had irritated the concessionaire, who insisted that he be set straight.

Concessionaires in the large, heavily visited parks have become very corporate and very powerful. They don't deal locally with park superintendents on issues that affect their income and profit. They go to Washington, but they don't fool around with officials of the National Park Service or the Interior Department. James M. Ridenour, who served as director of the National Park Service from 1989 to 1993 (under the first George Bush), wrote a book about his experience, *The National Park Service Compromised: Pork Barrel Politics and America's Treasures*. His revelations include an account of meeting with the Washington representatives of the Yosemite concessionaire. They were Robert Strauss, the former chairman of the Democratic National Committee, and Howard Baker, the former Republican leader in the U.S. Senate, who directly set him straight on their power and influence.

The Yosemite concession is part of the far-flung operation of Delaware North, an outfit that has outlived a shady past and is identified largely with operations at racetracks and baseball stadiums. It doesn't seem congruent to me to public service in a national park. From Tuolumne, my wife and I came down the mountain to spend a couple of days in Yosemite Valley. It was a noisy, congested, mechanized tourist ghetto, a money machine for Delaware North. In my speech I called it "more like a city than a park," but if we had gone there first I would have said more.

■   ■   ■   ■

National parks are the best of America, the final fragments of the original America as God made it. They embody history and art and the political process of democracy, all those values together, in what they protect and in how they came to be.

Consider that Thomas Moran, the artist, went to Yellowstone in 1871 with the official survey led by Ferdinand Hayden. Once that party returned East, Moran's watercolors and the photographs of William Henry Jackson were presented by Hayden at numerous congressional hearings as visible proof of Yellowstone's wonders, leading to a unanimous vote to preserve them in a national park. Two years later, at the request of Maj. John Wesley Powell, Moran went to the Grand Canyon. There he witnessed a raging thunderstorm and made sketches for his largest and most powerful painting, *The Chasm of the Colorado.* On completion in 1874 the painting was purchased by Congress to hang in the Senate lobby opposite its companion piece, Moran's earlier masterwork, the *Grand Canyon of the Yellowstone.*

I wish that those in Congress responsible for policies governing the parks, particularly those who currently appear intent on disassembling our national park system and closing parks as though they were obsolete military bases, would revisit those major works of art at the Capitol, feel the spirit of Earth and history preserved and protected, and set their sights on higher goals.

They may profess a desire to undo mistakes of the past of their predecessors. Certainly Congress at times has been known to bend the rules in designating areas not wholly qualified—it's called "bringing home the pork" or "park barrel" politics. Somehow, however, each and every one of these places takes on a special quality. I never met a national park I didn't like, or that I didn't think told an important story about America.

Trouble is, if you ask me, that national parks, despite their immense popularity, are too much taken for granted, first by the public and consequently by Congress. Members of Congress generally are not ecologists, or historians, or scientists, by education or experience, and only occasionally so by inclination. Politics they know better, and they don't see national parks high on the agenda, like prisons, taxes, and Medicare.

The time is right for a new campaign to bring national parks to the front burner, to show, in a thoroughly nonpartisan way, that national parks are essential to national well being. History is all on the side of the parks, with preeminent Republicans and Democrats, conservatives and liberals alike, as their champions down through the years. For example, when Theodore Roosevelt came to California in 1903 he strongly supported preserving both the coast redwoods and the giant sequoias of the Sierra Nevada: "I feel most emphatically that we should not turn into shingles a tree which was old when the first Egyptian conqueror penetrated to the valley of the Euphrates, which it has taken so many thousands of years to build up, and which can be put to better use. That, you may say, is not looking at the matter from the practical standpoint. There is nothing more practical than the preservation of beauty, than the preservation of anything that appeals to the higher emotions of mankind."

Then years later Franklin Roosevelt said, "I see an America whose rivers and valleys and lakes—hills and streams and plains—the mountains over our land and nature's wealth under the sea—are protected as the rightful heritage of the people."

I will also cite the crusading work in the Northwest of Irving Brant, who turned his talent in journalism to the preservation cause. In the monograph *The Olympic Forests for a National Park,* published by the Emergency Conservation Committee in 1938, Brant wrote these inspiring words: "Let this land, which belongs to the American people, be placed beyond the despoiling axe and saw, beyond the hunter's rifle, and we shall have for our own enjoyment, and shall hand down to posterity, something better than an indestructible mountain surrounded by a wilderness of stumps."

Enos Mills, the Colorado disciple of John Muir, summed it all up, for his generation and for all generations. "Without parks and outdoor life," he wrote, "all that is best in civilization will be smothered. To save ourselves—to enable us to live at our best and happiest—parks are necessary."

The early activists appreciated national parks everywhere, but focused politically on the particular parks in their home areas. Indeed, the entire history of the national parks and of the National Park Service is a record of democracy at work. Establishment of the agency in 1916 and of individual parks were achieved through the efforts of concerned Americans. They worked closely with professionals in government, but citizen responsibility has always been the vital measure of difference.

Virtually every one of the parks that we now take for granted came about because somewhere out there people cared, individuals who organized groups and then campaigned through the political process, without regard for any personal reward. They were patriots, of the best kind, and the same is true of park protagonists in our time.

I look back to the 1950s, when the proposed construction of dams in Dinosaur National Monument, though remote and little known, was defeated by a nationwide citizen campaign; to the expansion of the national park system to include lands formerly commercially exploited, in the Redwoods and North Cascades; to the successful campaign for Golden Gate in and around San Francisco, and for the great parks in Alaska; to the latest areas added to the system and worthy others proposed for it. Manzanar National Historic Site in California commemorates the unjust imprisonment of Japanese Americans during World War II, a reminder of the eternal vigilance needed to sustain civil rights and make democracy work. Fort Jefferson National Monument, off the tip of Florida in the Gulf of Mexico, has been reclassified and renamed Dry Tortugas National Park to reflect significance of the least disturbed coral reef system in the United States. Hopefully we will have a new national historical park protecting Civil War battlefields in the Shenandoah Valley of Virginia.

With passage of the California Desert Protection Act in 1994, Congress established two major new national parks, Death Valley and Joshua Tree, and the Mojave National Preserve. The action was taken in response to rising awareness of desert beauty, history, science, and wildlife, coupled with concern over intensive exploitation and damage. The Mojave has been the most contested, but perseverance in its behalf ultimately will succeed.

It takes serious commitment to make such things happen. But the effort in itself is rewarding, even more than whatever success the effort may bring. Involvement evokes the best in people. I love the work that individuals do, rising above themselves, and above institutions.

Democracy starts with community action, scarcely ever with answers from above. True enough, the environmental crisis is global, which means that government—along with universities, media, science, and organizations—have their place and purpose, but the primary role belongs to people exercising the rights and responsibilities of citizenship. Community involvement enables people to analyze problems, test ideas, learn from experience of others, and feel empowered.

Working to make a difference helps to understand that efforts and energies are rewarded, even in losing causes, or causes that appear lost. In Memphis, the Citizens to Preserve Overton Park during the 1960s and

1970s were determined to save one of the finest urban forests in the world from proposed construction of a highway through the middle of it. They were forced to contest not only merchants, developers, and public officials but also the two powerful Memphis daily newspapers, which ridiculed park defenders and belittled any politician who dared to speak in defense of the park. The parks department and park professionals acquiesced to the planned highway, but the citizens group insisted that an established park represents an integral and sacred part of the American city, that it makes the city habitable, and that it would make more sense to locate the highway elsewhere or not build it at all. The Overton Park case, because it involved federal highway funds, was debated in Congress and before the Supreme Court. "The park may be lost," said a leader of the citizens group, "but our lives are enriched for each day that we save it." Ultimately, Overton was spared, and it still enriches the life of Memphis.

People are ready to do these things. Students in a course I taught last year taught me a thing or two about awareness and involvement. They responded to an assignment with thoughtful, positive ideas. One wrote, "The desire to gain knowledge is the first step toward making a difference." And another, "The courageous first step begins with both the development of an individual's true sense of human dignity and the identification of humanity's place in nature." A third wrote, "We have to keep fighting and keep a positive attitude in order to make a change," citing Whitman: "Faith is the antiseptic of the soul."

Others wrote that people need to start conserving in their own communities, that if we start making a difference community by community, then hopefully in time the scale will expand to a larger, global attitude. Then there was this comment:

> Environmental awareness has rapidly increased, not only nationwide but worldwide as well. While people continue to harm the environment, many are changing their ways and taking action to protect our natural surroundings.
>
> I have seen evidence of increased awareness of our ecological problems in my everyday life. Most people I know recycle cans, glass, plastic bags and newspaper.
>
> As a child I remember going to picnics and parties and throwing away my paper plates, napkins and cups.

I would use about five different paper cups at the same event, and it didn't bother me or anyone else. I recall seeing my next-door neighbor pouring oil down the sewage drain. I didn't see anything wrong with that either. I would hear the word "extinction" but the only thing I would think of was the cute squirrel T-shirt on which I once saw the word. Extinction couldn't be that awful, if all the grownups were allowing it to happen.

Now I see a completely different type of child. She understands the importance of conserving and recycling. She knows the definitions of air pollution, greenhouse effect and ozone pollution.

Papers like that don't come along all the time, but they're no freak either.

If the young show this concern, adults should be with them. I wish I could say that our friends in the National Park Service provide the leadership, but the principles of Stephen Mather, Horace Albright and Newton Drury are not much practiced. The early park pioneers, both in and out of government, strived to protect and preserve the nation's treasures. That was their goal. They had no way of anticipating the cataclysmic changes in society—in population, transportation, the power of concessionaires, the emergence of a vast outdoor recreational industry, the advent of off-road, "all-terrain" vehicles (as Barry Goldwater calls them, "the Japanese revenge"), the transformation of gateway communities into tourist ghettos.

Unfortunately, the National Park Service as an institution with a cause is only a shadow of itself. Many administrators find it convenient to support, or propose, construction projects designed to draw visitors and to sanction crowd-pleasing activities reduced to a low common denominator rather than to focus on the values of the resource itself and its protection.

Thus, more than seventy-five thousand snowmobiles descend on Yellowstone during the winter. Presumably they are limited to the unplowed roads, but there is little ranger supervision or control. As the numbers continue to rise, so does the pressure to open more of the park to them. That is our so-called "flagship" national park. Tight little Yosemite Valley, with thirty thousand visitors daily, is more like a city than a park. Park personnel call these places "sacrifice areas," as though to legitimize the sacrifice. John Muir said, "Come to the mountains, for here there is rest." He didn't say it would be in a lodging facility with bath, bar, restaurant, entertainment, bike rentals, ice skating, and room service. If that valley were relieved of one-half of its buildings, automobiles, and people, it would become twice the national park it is today.

In my agenda there should be fewer roads in the parks, and no off-road vehicles, no airplane or helicopter sightseeing disruptions of pioneering adventure. National parks should not be reduced to popcorn playgrounds or theme parks but preserved as sanctuaries, sustaining the soul of America; they are priceless time capsules for tomorrow that we are privileged to know and enjoy today. Chronic overuse and misuse—"loving the parks to death"—clearly deplete the visible physical resource that people care about, disrupting the spiritual connection between the visitor and something larger than himself or herself, and more lasting than all the mechanization of life and work at home.

Theodore Roosevelt called the Grand Canyon "one of the great sights which every American if he can travel at all should see," but he didn't ask everyone to come at the same time. To the contrary, he pleaded with Americans to do nothing to mar the canyon's grandeur. That should be our goal.

National parks when they were new yielded discovery, adventure, and challenge. They should always do that. And they can, if we set our minds and hearts to it. As the rest of the country becomes developed, and supercivilized, national parks should be held apart, safeguarded to represent another side of America, free of technology, free of commerce and crowds, free of instant gratification, a pioneer, self-reliant side of America.

In the unending struggle to obtain and sustain a quality environment of life, national parks contribute by their existence as model ecosystems. The challenge is to raise and then sustain the quality of the parks themselves. Turning over the National Park System in better condition than we found it may be the most important bequest of this generation to the future.

No More "Sacrifice Areas," Please

At the Forum on the Future of the Blue Ridge Parkway,
Asheville, North Carolina, October 5, 1998.

A bonus of the trip to Asheville was a side excursion to Flat Rock, fifteen or twenty miles away, to visit Connemara, the home and farm
where Carl Sandburg spent the last twenty-two years of his life. It was too
good and too close to miss. Thanks to designation as a national historic
site, the property remains largely as Sandburg and his wife left it, reflecting his immersion in poetry and history and hers as a breeder of prize
goats. At the gift shop, I bought a copy of *The People, Yes* and read into
it that night. It reminded me that Sandburg was really a journalist who
collected Americana and chronicled Americans in his own way.

Thoughts turned from Sandburg to Walt Whitman. He too started as
a journalist, to become in time the prophet of liberty through poetry of his
own structure. He published the first edition of *Leaves of Grass* at his own
expense and never wavered from his chosen course. Once I visited and
wrote about the only home he ever owned, the two-story clapboard
house in Camden, New Jersey, which he called his "little shanty." It was clut-
tered and rundown in a rundown neighborhood but it was pure Whitman.
I wish that the preservation of his home and interpretation of life and
work could properly honor him, as Sandburg is honored at Flat Rock.

■　■　■　■

Over the past half century I have witnessed many, many changes in the
national parks, some few for the better, but others highly damaging and
cause for serious concern. These precious places are overused, misused,
polluted, inadequately protected, and unmercifully exploited commercially and politically. We the people need to redefine and reassert the

rightful role of national parks in the fabric of contemporary high-tech, materialist-minded, profit-driven society. We need to rescue the national parks from being reduced to popcorn playgrounds.

The Blue Ridge Parkway is no exception. I appreciate the concern of citizens here for the fate and future of the parkway and its environs. I applaud the citizen organizations for their involvement and for raising questions about incompatible developments along the parkway boundaries, the ongoing construction of a substantial headquarters complex evidently without public input, and plans to construct an Imax theater, diverting the public from the genuine article to a synthetic commercial substitute. Whatever the issues, people who care belong at the table and should feel welcome by park administrators.

I feel honored to be here, for my interest and activity in national parks really began in the southern mountains. I made my first trip to the Great Smokies in 1946. I first came to the Blue Ridge Parkway in the 1950s while I lived in northern Virginia, outside of Washington, D.C., and was developing my career as a travel writer. My files show that I wrote my first article about the parkway in the *New York Herald Tribune* in April 1952. I recorded that with newly paved sections a total of 320 miles of the parkway would shortly be completed, out of the ultimate total of 468 miles between Waynesboro, Virginia, and Soco Gap, North Carolina, the gateway to the Smokies.

I knew little about conservation. My interest was in telling readers of the travel pages about some special place worth visiting. And that was what I found here. I quoted the park superintendent, Sam P. Weems: "The parkway is the longest road in the country—and probably in the world—designed solely for recreational travel, free of commercial sign-posting and traffic." I wrote further that together with the two national parks, which it links—Great Smoky Mountains of North Carolina–Tennessee and Shenandoah of Virginia—the Blue Ridge Parkway comprises the last large natural frontier in the eastern United States. I described the views from scenic overlooks and the footpaths in primeval forests and on mountain crests and the vestiges of old human culture. "Besides its natural attractions," I wrote, "the Blue Ridge is a treasure house of mountain folklore, preserving isolated old cabin homes and mills that reflect the lives and habits of rugged highlands pioneers."

Over the years I traveled around the country, exploring national parks everywhere, and for many years I kept apace by writing and updating the annual *Rand McNally National Park Guide*. But repeatedly I returned here, hiking, camping, driving the parkway, entering new crannies, coves, and hollows like Gingercake Mountain, Dogback Mountain, Sitting Bear, Hawksbill, Table Rock, and Devils Hole Branch. In the early 1960s I came to research and write my book *Strangers in High Places: The Story of the Great Smoky Mountains*. I brought my family for a while and my children attended Sunday school at the Macedonia Baptist Church in Cherokee, where lessons were conducted both in English and the native tongue. I admired the Cherokee in many ways. They were great crafters, the women especially, collecting river cane, split oak, and honeysuckle and painstakingly developing dyes from roots and leaves to make beautiful baskets that would make their grandmothers proud.

In due course it struck me that the abundant and diverse plants, flowers, and trees are to the Smokies and Blue Ridge what granite domes are to Yosemite, geysers to Yellowstone, and the wide, deep chasm to the Grand Canyon. I consider the Blue Ridge Parkway a public treasure if ever there was one, more than a scenic roadway but a composition of unspoiled natural, historical, and recreational areas, as valuable as any of the great western areas.

I became friends with Stanley Abbott, Sam P. Weems, and Granville Liles, early superintendents of the Blue Ridge Parkway, all of whom were committed to the highest standards of public service. I met many fine people in public agencies and private enterprises who seemed to me working with common spirit to conserve the treasures of the region while making them available for wholesome recreation. The wildflower pilgrimages and handicraft fairs were major events for tourists and locals alike.

In those days tourism was more like a cottage industry, wholly compatible with conservation and national park purposes. Now, in contrast, industrial tourism has become an overpowering, politically dominating influence. In the prevailing politics of our time a national park is considered valid or defensible as long as it helps jingle the cash registers of local merchants and tour companies. Urbanites are made to feel comfortable in the backcountry with treeless, barren camping suburbias. Congestion, noise, the intrusion of mechanistic supercivilization interfere with qualities that make the parks special.

The heart of Yellowstone, our oldest national park, has been reduced to an urban tourist ghetto, complete with crime, litter, defacement, and vandalism. In California, Yosemite Valley is more like a city than a park away from cities. The proposed management plan for Glacier National Park, the heart of the grizzly-critical ecosystem in northern

Montana, changes "natural zone" to "visitor services zones." The entire Going to the Sun corridor would be changed from natural and historic to "visitor services," thus allowing future congested motorized boating at Lakes McDonald and St. Mary. "Wilderness" as a classification has been removed and is not mentioned or defined, thus opening the way for the proposed new winterized motel, fast food restaurant, and new parking lot. Park personnel call these places "sacrifice areas," as though to legitimize the sacrifice.

These examples are not the exception, they are the rule. National parks in our time are being converted to popcorn playgrounds, resource commodities for the benefit of for-profit park concessionaires, tour companies, and business interests in park-bordering communities like Gatlinburg, Tennessee; Jackson, Wyoming; and Moab, Utah, that have paid a heavy price in community character and quality of life.

I fear much the same prevails along the Blue Ridge Parkway now. For one thing, the magic mountain vistas have been diminished by smoke and haze, the result of toxic emissions airborne from distant factories and from power plants and automobiles. High forests have become catch basins for acid rain, seriously weakening stands of balsam, Fraser fir, and spruce that visitors come to enjoy. For another, condominiums and commercial developments of one kind and another have emerged on private lands bordering the Blue Ridge Parkway. But I do not hear expressions of concern from park administrators. The communities have changed from friendly mountain towns to hard-edged tourist entrepreneurships. At Cherokee, the gambling casino has become a much larger landmark than the Qualla crafts shop. On the parkway, the concessions have grown bigger but not necessarily better. And the biggest new thing, the parkway headquarters cum tourist center is built on ecologically sensitive terrain, with more response to tourism than to ecology. The unfortunate truth is that many administrators find it politically convenient and advantageous to support, or propose, construction of such projects.

Politics and political pressures are not new, nor necessarily evil. The entire history of the national parks and of the National Park Service is a record of the political process at work. The people of North Carolina and Tennessee know that well, for there would not be a Great Smoky Mountains National Park without their concerted interest and support. Without nationwide citizen campaigns in the 1960s there would be no national parks in the Redwoods of California or the North Cascades of Washington State. Were it not for caring citizens, the Colorado River would be dammed where it runs through the Grand Canyon, the great forests would be long gone from the Olympic Peninsula, and the Great Smoky Mountains would be scarred with a transmountain highway.

I well remember coming here in 1966, when the National Park Service chose the Great Smoky Mountains for its first national park wilderness. History records the incredible: Instead of pressing the cause of wilderness, Director George B. Hartzog Jr. personally pushed for a multi-million-dollar transmountain road across the park, plus additional inner loops and massive campgrounds. It was something he conceived in the back rooms of political power, without public consultation. He and his agency were motivated not by desire to protect wilderness but to weaken it, so that parks would remain open to mass recreational and commercial development. If ever there was a time for citizens to be heard, that was it—and were they ever!

"It is amazing how many persons from all over the country supported wilderness designation," Ernest Dickerman, a leader in the citizen Save-the-Smokies crusade, wrote to me in retrospect thirty years later, "and opposed any new roads in the course of the campaign— which lasted six years from 1965 until 1971, when George Hartzog finally threw in the towel. Frankly, the Park Service, except perhaps during its earliest years, has commonly been out of touch with the owners of the national parks in its basic policies and practices. The Park Service, instead of working closely with the citizens knowledgeable about national parks and devoted to protecting their extraordinary natural values, has considered them as antagonists."

In contrast to Hartzog and others of recent times, Director Newton B. Drury during World War II resisted pressures to open the parks for military purposes. Consequently, little damage was done, none needlessly. Following the war, new demands arose to open the parks to mining, logging, grazing, and dams. Drury held firm, insisting, "If we are going to succeed in preserving the greatness of the national parks, they must be held inviolate. They represent the last stands of primitive America. If we are going to whittle away at them we should recognize, at the very beginning, that all such whittlings are cumulative and that the end result will be mediocrity."

I've heard it said many ways but scarcely said better—one simple small exception may not hurt, but making exceptions gets to be a nasty, destructive habit. Stanley Abbott once said to me that he was "a Drury man." It was the Drury principle that he followed in pioneering administration of the Blue Ridge Parkway. We should halt the whittling and refuse to accept mediocrity.

We need to reach out to find allies. I hope this forum will prove the beginning of it in the southern mountains. Whatever the issue, people who care belong at the table and should feel welcome by park administrators. One park superintendent recently told me, "After 27 years in the NPS, I have been imbued with the genius of management, and therefore what I do to this park is what is best for this park." That won't do and we should not accept it.

It isn't easy to be heard. Sometimes the challenge seems impossible. Yet the history of our national parks, like the history of our country, documents miracles by people who cared. Plainly, those who work to safeguard the Blue Ridge Parkway as a cathedral of hope will be blessed.

I Prayed to Rise above Paradigms, Postulates, Phenomenology, and Epistemology

Accepting the award as Outstanding Alumnus of the Year,
at the Presidential Awards Dinner, the Union Institute,
Cincinnati, Ohio, September 24, 1999.

Since I was getting into a doctoral program, I read with interest the book *Killing the Spirit: Higher Education in America,* by Page Smith, in which he details oppression in the training and employment of Ph.Ds.

Then I also read *With Justice for None: Destroying an American Myth,* by Gerry Spence, the famed trial lawyer, who asserts that prestigious law schools prepare their students to work in prestigious law firms wholly committed to serving corporate America. As Spence sees it, citizens without wealth or power rarely receive justice. And maybe his criticism applies not only to law schools but also to the whole system of education.

Happily, I did not feel that way about the doctoral program I went through. It wasn't oppressive, it was liberating. I felt it was meant to train educated people to serve citizens without wealth or power.

■　■　■　■

At the outset of doctoral studies in 1991 I wrote in my diary that if I could be myself going into the program I would be myself coming out of it. I wanted to learn the lessons without being overwhelmed by methodologies and modalities, paradigms, postulates, phenomenology, and epistemology. I prayed to be the best at what I am: to be a good journalist, who regards the work as not work but as a calling, a means of changing society for the better, who conducts thorough research, subjects data to critical analysis, and expresses a strong viewpoint clearly, without fear or favor.

I believed then, as I do now, that even while emotion is the most powerful force in human life, the training of professionals generally represses emotion. I wanted to prove it doesn't have to be that way; I wanted to legitimize subjectivity, the expression of heartfelt personal viewpoint in learning as well as in journalism. That desire was an important factor in my decision to apply for admission to the Union Institute.

Initially I was interested in studying the lives and work of journalists and writers of the 1940s, 1950s, and 1960s who committed their talents to wilderness and conservation, leading the way to passage of the Wilderness Act of 1964. Ultimately I decided to leave that for later. Then, once I settled on my project, I expected to cover a lot more ground than was either necessary or feasible; I thought I would include everything from the history of the wilderness movement to a critique of performance by federal agencies responsible for wilderness administration.

The members of my doctoral committee helped me a great deal. All of them wanted me to do the best work I could. They guided me to identify very specific areas of study. Their incisive critiques of my draft essays sparked fresh learning; they reminded me to heed the call of social relevance. I appreciated their respect for my role as chair, but even more their respect for each other and their sense of responsibility for the work of the committee as a whole. Following the pregraduation meeting, Dr. Liane Russell said with a sigh and a smile, "I'm sorry it's over." I might have said the same, though, happily, the connections both personal and professional have lasted.

The design of the program was of my own making, of what I wanted to do, with flexibility to alter emphasis where it would help. I attained my goal of the degree in two years of focused learning, even while doing other things too. I could do them because they were worthwhile and enriching. And still there is so much to do.

The support and encouragement of scholars, and of administrative personnel at the Union Institute, meant a great deal. Undertaking a serious doctoral program plainly is a major venture, a test of discipline as well as intellect, particularly in the Union Institute system of self-design and self-responsibility. The learner needs all the assurance and reassurance he or she can find; I know that I did. Luckily, I was blessed by the words and thoughts of members of the Union faculty, notably my advisors, Ben Davis and John Tallmadge, and Penny MacElveen-Hoehn, who wrote to me following the entry colloquium (of which she had been co-convenor). The colloquium had been a time of stress, with twenty new people, highly charged, sailing into uncertain waters, on the same ship, yet heading on different routes. But Penny wrote to me as follows: "I feel confident you will hold on to your ability to communicate with power and clarity in spite of doctoral level education. Also, you've been around long enough to know that you have resisted many forces without surrendering your passion and that you won't lose it to the Union! I have great respect for your vitality (intellectual, physical and spiritual) and ability to commit to this new enterprise for your life."

I read and reread that letter many times along the journey. It told me to be of good cheer because I could reach my destination, and that I would by keeping faith with myself and by trusting the process.

My goal now is to build upon experience and studies to look broader and deeper at the environment. I want to relate the environment to other social issues and to depersonalizing institutionalism, for they derive from the same roots in a society that critically needs to examine itself and change direction. The real questions for me are, How can those who hope to heal the earth join with those who hope to heal the soul? How can wilderness preservation encourage an attitude with spiritual ecological dimensions?

Courage, honesty, and a sense of purpose are essential for the individual who wants to face the world with hope and heart to make it whole again. Conservation, as I know it, responds to social needs. It treats ecology as the economics of nature, in a manner directly related to the economics of humankind. Keeping biotic diversity alive, for example, is the surest means of keeping humanity alive. But conservation transcends economics—it illuminates the human condition by refusing to put a price tag on the priceless. I want to help define the human condition in subjective values essential to personal transformation as the starting point to transformation of global society.

I found myself so thoroughly absorbed in the Union Institute process that I had little time or inclination for criticism. I had too much to do and encountered too many wonderful and helpful people to find faults. Besides, I learned an important lesson from a fellow learner that I will relate in closing.

As part of my program I attended a seminar on music appreciation at the University of Indiana. For almost a week I attended concerts, recitals, and lectures at Indiana's famous School of Music and enjoyed them all. But I wondered if I was indulging more in pleasure than education. As in all cases, the student gets back in direct ratio to what he or she puts into it. Robert Palumbo, a fellow learner, working with clients who are chemically dependent and afflicted with AIDS, showed how he raised the experience to a higher level: At coffee the last day of the seminar he told me of how much he had learned relative to music in healing, particularly from a book he found at the university library published in Italian, no less, on *Music and the Human Body.*

It is impossible for me to quantify the value of this brief unstructured conversation, but it symbolizes the best of the experience, of feeling the positive influence of people and of exploring new sources of knowledge while working alone and testing myself day after day. Best of all, the Ph.D. marks not the end, but a beginning.

Forty and Fifty Years Ago
They Were Missionaries

At the Northwest Wilderness Conference,
Seattle, April 1, 2000.

David Brower was plainly pretty weak when he came to speak at the wilderness conference in Seattle in April 2000, barely eight months before he died. He walked with a cane and needed help getting up to the platform, where he sat with Celia Hunter of Alaska, Polly Dyer of Seattle, and me. He was the one the crowd came to see and hear; we were merely the supporting cast. He looked his years, eighty-eight, but when he spoke he was as strong as ever—strong in commitment, powerful in inspiration. Brower has been called the most important conservationist since John Muir, and that may be so. He was always creative and daring, no mountain too high to climb, no battle of principle too tough to fight.

I think of Dave Brower speaking as executive director of the Sierra Club at a forum at the Grand Canyon in 1967 and of the full-page advertisement he ran for the club in the *New York Times* comparing flooding the Grand Canyon with flooding the Sistine Chapel. He simply wasn't going to let that happen. Secretary of the Interior Stewart L. Udall, who supported the proposed dams, later said, "The Sierra Club didn't save the Grand Canyon, the American people did." But it was Brower who lit the torch and mobilized America.

In 1957 he helped a handful of Washington state activists organize the North Cascades Conservation Council. "In the early 1960s, as it grew obvious we locals had to 'go national,'" Harvey Manning later recorded, "Dave's leadership became paramount. He knew all buttons of all the players in the national game. He pushed them. And thus, in 1968 was created the North Cascades National Park."

Brower was gifted artistically and might have been successful as author, editor, or filmmaker. To advance the North Cascades cause he produced and personally photographed a beautiful documentary film, *The Alps of Stehekin.* He launched the distinctive Sierra Club Format Books with *This Is the American Earth,* written by Nancy Newhall and illustrated by Ansel Adams. Commercial publishers later copied the style, but they produced mere coffee-table books.

While doing such creative things with one hand, in 1966 with the other hand Brower hired Gordon Robinson as the Sierra Club's first professional forester. Until then the irate public had felt ill-equipped and inadequate to debate the Forest Service and timber industry over technical issues of "allowable cut effects" and "mean annual increments." Robinson was able to address these issues in a professional manner. He was a graduate forester (University of California, Berkeley), who had worked for many years managing large blocks of commercial timberland for the Southern Pacific Railroad, the largest private landholder in California. When he joined the Sierra Club staff, Robinson confirmed suspicions that something terribly wrong was taking place. Now at last the club and the environmental movement could decode the jargon, penetrate the curtain of professional expertise, and even the scales in debate.

Brower's genius lay in his clarity of purpose, but he came across at times as reserved, or aloof, perhaps arrogant. He made important commitments without approval of the Sierra Club board of directors, which ultimately chose to depose him. In 1969 he established Friends of the Earth, and later, when time came to move on from there, Earth Island Institute.

Stewart Brandborg, at the Wilderness Society, was another environmental leader who suffered dismissal. In the years I spent in Washington, D.C., Brandborg was my best source and best mentor. He answered many, many questions with patience and good humor, in contrast with others whom I found absorbed with self-importance, snobbery, and jealousy of their peers. I have never, to this day, fathomed mean-spiritedness among people presumably working toward common goals in the public interest. With Brandborg, however, I observed that both his door and heart were open to many different kinds of people, whether from the *New York Times* or a local citizen group from Idaho or Iowa.

Brandborg served for fifteen years as executive director of the Wilderness Society, following Howard Zahniser, who died in 1964, a few months before the Wilderness Act (which Zahniser had drafted) became law. For Brandborg the act opened the way to a new level of citizen involvement and activism, and he committed himself to the grass-roots conservation movement. But his dismissal in 1976 was unhappy and acrimonious. This was explained later in a very small part of a very long article on the history of the Wilderness Society published in its magazine. The author downplayed Brandborg's role in the fight for wilderness, asserting he functioned "at too fevered an emotional pitch." Being treated as a nonperson without any mention would have been kinder.

Paul Pritchard, another friend, met a somewhat similar fate at the National Parks and Conservation Association. Pritchard had worked as a planner in the Georgia state government when Jimmy Carter was governor, then as an official of the Interior Department in Washington. When he took over the NPCA in 1980, it was on shaky grounds with about twenty-five thousand members. When he left eighteen years later, it had five hundred thousand members. Although many were attracted by mail-order promotions, they provided the resources to build an organization. Pritchard established a field staff around the country that made NPCA a presence of influence and conducted forums in Washington and elsewhere on the problems and future of the parks.

I remember one at a resort just outside Grand Teton National Park, Wyoming, in September 1981, designed "to bring together concerned citizen experts to provide a framework for the development of a comprehensive National Park System plan and, second, to provide a basis for alerting the public about problems facing the national parks." It was a very difficult time for the national parks, with James G. Watt and other Reaganauts running, and ruining, the Interior Department, so this conference at least gave them reason to pause.

The NPCA under Pritchard was not especially militant. It lacked the crusading edge that Brower manifest wherever he went. Pritchard nevertheless provided a forum; he certainly encouraged me. But some of his directors felt he should have been more in harmony with the National Park Service and in 1998 he was booted out unceremoniously. Then he fooled the directors by taking over the National Park Trust, raising money to acquire private land for park purposes.

Forty or fifty years ago they were missionaries. In the years since then I watched leaders of national environmental organizations change from missionaries to corporate CEOs. David Brower came back to the Sierra Club as a member of the board of directors but in May 2000 resigned out of frustration. "The world is burning and all I hear from them is the music of violins," he declared. "The planet is being trashed, but the board has no real sense of urgency. We need to try to save the earth at least as fast as it's being destroyed."

In my book he was right as rain: We still are subject to the music of violins. In January 2001, when President George W. Bush nominated Gail Norton, a protégée of James G. Watt, with a dismal environmental record of her own, to be secretary of the interior, many citizen groups urged the Senate not to confirm. However, several said, "Well, give her chance." These included the National Audubon Society, World Wildlife Fund, Environmental Defense, National Wildlife Federation, and National Parks and Conservation Association. Thomas Kiernan of NPCA (who succeeded Pritchard) declared, "To label Norton as either anti-park or an ally of the parks would be presumptuous at present. Although her history on wider environmental issues gives us profound concern, she has the opportunity to lead on park issues."

That is precisely the type of statement I learned to expect from the leader of an environmental organization that has turned tame, corporate, and compromising, replacing activism with pragmatic politics. And after I delivered my speech at the Northwest Wilderness Conference, reprinted below, and the floor was opened for questions, the president of the Wilderness Society, William Meadows, rose from the audience to set me straight. He was angry and upset while exclaiming that I did not understand all the good and important works performed by the Wilderness Society. Then, when the session adjourned, and a crowd of people gathered to talk with me at the podium, Meadows came forward and tried to shove them all aside to impress upon me the validity of the new management plan for Yosemite National Park. Perhaps in some other circumstance we might have discussed it rationally, but here he was anything but convincing.

But there is more to it. Traditional concepts of environmentalism overlook hidden realities—unpleasant social issues and problems that environmental advocates find convenient to avoid. This struck me when I read in manuscript the opening page of *The Boys on the Porch,* a book by June Eastvold, my wife. She wrote that principles of justice, compassion, and transformation have been clearly defined in the three words "Love your neighbor," but then asked how such a simple statement can be so difficult to put into practice:

> Studies on social problems and class struggles point
> to the need for change. Yet, despite volumes of data, things

stay the same. People caught in poverty remain on the margin. Middle and upper class citizens intentionally construct a daily lifestyle that eliminates direct contact with the poor. The extremes have no meeting ground.

One institution charged with providing a community of mutual acceptance is the church. Yet racial and class separation remain strong there too.

It isn't only in her church where these separations remain strong. It's my church too. I never see blacks at environmental conferences. I never hear a Chicano rise at a hearing to say, "I represent the local Sierra Club chapter." When I read the quarterly journal of the Natural Resource Defense Council I find a list of board members and their affiliations. They are largely Wall Street lawyers and investment counselors, and the same is true of other national environmental organizations. Poor people are not represented on their governing boards, except as an occasional token.

But, then, racial and class separation remain strong all across society. Homeless people sleeping in doorways on the streets are taken for granted. So are children playing in streets surrounded by smokestacks, toxic sites, incinerators, and freeways, living with pollution, noise, drugs, disease, and chaos. They are low income, black and Latino, people of color pushed into areas unfit for living because they can't afford better housing. And they are invisible.

I've learned that if we can't save the environment for them, where they live, then we can't save it anywhere else either. The seemingly distant national parks and wilderness areas are merely on loan until needed by the corporate system. Alfred Runte, the historian, developed his "worthless lands theory," asserting that corporate power allowed establishment of national parks because the areas involved were not good for anything else. But environmental groups and their leaders don't want to hear that. They are so busy with wildland ecology they generously ignore urban ecology, even though the forces that threaten wilderness threaten the places where people live. They do not deal with racism, violence, and injustice, forces that exploit many, many people, human resources, in circumstances where they live.

The Northwest Ecosystem Alliance is headquartered in Bellingham, Washington, where I live. It raised $16 million (with large chunks from computer tycoons in Seattle) to purchase logging rights in the Loomis state forest, adjacent to the Pasayten wilderness in eastern Washington. My neighbors and I turned to the Northwest Ecosystem Alliance for support in fighting a critical issue in our own backyard. We thought we had a case.

Bellingham does not retain its natural green setting. It is losing open green space in every part of the city. It does not protect unique natural features. If it did, the city would not have approved a massive, ill-conceived, 172-unit subdivision in a heavily forested area called Park Ridge. It was clear at the very beginning of the proposal that the site is fraught with obstacles to development, including two streams running through the full length of the property, numerous extreme slopes and ridges, and limited traffic capacity. Consequently we neighbors organized the Concerned Citizens of Park Ridge. We wrote letters, testified at countless hearings, and raised funds to bring legal action. We sought the cooperation and support of the Northwest Ecosystem Alliance but repeatedly were brushed off because *"this is only a small local neighborhood issue."*

The Northwest Ecosystem Alliance organized the Cascades Conservation Partnership and initiated a fundraising campaign to save forests in "the fragile central Cascades ecosystem . . . and preserve critical habitat for gray wolf, grizzly bear, pine marten, endangered salmon runs, and other wildlife." But all types of animals exist in all types of places. Raccoons, squirrels, mice, opossums, pigeons, and skunks make urban areas incredibly wild and natural in their own right. Our eighty-acre Park Ridge embraces lush vegetation, wetlands, and a thriving, diverse wildlife population, remarkably within the city limits of Bellingham. Park Ridge provides a growing community with green lungs, opportunities for outdoor recreation and environmental education to complement classroom studies.

If you ask me, we need a new, inclusive environmental vision, a world view of nature and society that brings concerns and interests of all brethren and sisters, human and nonhuman animals, into our moral gaze.

■ ■ ■ ■

In 1956 Representative John P. Saylor of Pennsylvania introduced the Wilderness Bill in the House of Representatives. For eight years Saylor led the uphill legislative battle and never gave up. In 1961, when the going was tough, he declared, "I cannot believe the American people have become so crass, so dollar-minded, so exploitation-conscious that they must develop every last little bit of wilderness that still exists."

John Saylor was by any measure a wilderness hero. Personally, I have never thought of myself as particularly heroic in the wilderness movement, despite the flattering title of the program here, "An Evening with Conservation Greats." The best I can say for myself is that I was on the scene during a critical and exciting period, and that I was privileged to know the principals like Saylor and to write with caring about their activities and the issues.

Harvey Broome, one of the founders of the Wilderness Society and president of the society when the Wilderness Act became law, was a special hero and friend of mine. Shortly after he died in 1968, Representative Saylor paid tribute to Harvey on the House floor with these words: "We must resolve never to falter, as he never faltered, and to take inspiration from his life to fight all the harder for the future of the wilderness. His spirit knows no boundaries and will be with us in the years ahead."

I think that's what it's all about and why we're here today. Those people of forty and fifty years ago were missionaries and visionaries, giving broad shoulders to stand on. Howard Zahniser, who drafted the Wilderness bill, created reality out of a dream. I remember many beautiful people like Zahnie, Harvey Broome, and John Saylor, who supported the cause. For example, the National Audubon Society was headed by Carl Bucheister and Charles Callison, who were motivated by principle and high purpose, and when he retired, Callison established the Public Lands Institute with his own money and energy to defend the integrity of the people's landed estate. Oh Charlie, where are you when we need you?

But something has happened between then and now. In 1985 the National Audubon Society, Sierra Club, and Wilderness Society all were shopping for new executive directors. Those organizations were at midlife, facing marked transformation from volunteer efforts to business enterprise. They didn't look to grass-roots wilderness campaigners for their new executives. No, all three engaged professional search companies, specifying that they were looking for leaders strong in fund raising, finance and budget; they wanted management specialists—and were willing to pay high prices for them.

These and other national environmental organizations, I fear, have grown away from the grass roots to mirror the foxes they had been chasing. They seem to me to have turned tame, corporate, and compromising, into raging moderates replacing activism with pragmatic politics, and a willingness to settle for paper victories.

It grieves me deeply to read a statement by a Wilderness Society representative calling the new management plan for Yosemite National Park "an elegant balance between park protection and visitor use and enjoyment." It sickens me when this plan clearly would turn Yosemite

Valley into a pricey crowded commercial resort benefiting above all the park concessionaire, the multinational Delaware North, better known for its facilities at racetracks and baseball parks.

Likewise, it distresses me to learn that national environmental organizations have endorsed the proposed giveaway of 272 acres of the Kaibab National Forest at the gateway to Grand Canyon National Park for construction of 1,270 hotel rooms and 270,000 square feet of retail mall shopping, the equivalent of four large department stores. [The proposal was endorsed by the National Park Service and Forest Service but was soundly defeated in a statewide referendum in November 2000.]

Our public agencies, the Forest Service and National Park Service, have lost their way. They want to think of themselves as "marketers" of mass recreation as a commodity, building "partnerships" with commercial interests, the bigger the better, and treating the public as "customers." Environmentalists need to bring the agencies back on track as resource stewards in committed public service.

Providing sanctuary for America's wildlife heritage should be the single most important role of the national parks and wilderness of the national forests at a time when diversity on the planet is so thoroughly endangered. The National Park Service and Forest Service should be apostles and advocates for mountain lions, wolves, grizzly bears, and buffalo. Wild animals make a park a park, but wildlife has been crowded out of its habitat in every park without exception. Animals are not protected from snowmobiles, sightseeing airplanes, helicopters, tour buses, cars, concessionaires, hikers, bikers, and park administrators.

For this reason I support wholeheartedly the National Day of Action on June 10, initiated by Scott Silver of Wild Wilderness and cooperating grass-roots organizations to ensure that the voice of industrial recreation is not the only message heard by Congress and the media about the misguided, misleading and misanthropic Recreation Fee Demonstration Program. Stewardship of public lands—especially wilderness— often requires limitation of use, but fee demonstration provides a powerful incentive for managers to avoid anything that will limit use— the more use they can generate the greater their budgets. Money is not the simple answer, but Congress must provide the funding to do the necessary administration to maintain these national treasures for future generations. It should not order administrators to merchandise the resource in order to pay their salaries. That is the message to get across on June 10.

It will be a great day. I hope somehow that Butterfly Hill will be with us—that beautiful butterfly who lived on a platform six feet by eight feet in size from December 10, 1997, until December 18, 1999, high up in her beloved thousand-year-old tree in northern California. It was her island of hope in a sea of desolation. And when Butterfly climbed down, after 738 days, she said, "You can't protect animals without protecting their home, which is also our home."

Butterfly had a sign high in the tree that read Respect Your Elders, which leads me to tell her, on behalf of the heroes of the wilderness movement who are gone, Your elders respect you. John P. Saylor in 1968 said that he was proud to consider himself a fellow to Bob Marshall, Olaus Murie, Howard Zahniser, and Harvey Broome: "They were all great leaders," he said, "for the saving of wilderness for our time, for all time. They have passed on, but their legacy falls to new leaders, as their spirit lives on." Yes, their spirit lives on and the legacy is yours.

■   ■   ■   ■

Julia Butterfly Hill and Scott Silver, whom I mentioned separately at the close of this speech, represent irreverent new approaches to environmental issues, different from the civility characteristic of established national organizations. Butterfly acted alone but as part of a group of young nature defenders, tree-sitters, determined to save a critically endangered redwood forest in northern California. She set up camp on a slender platform 180 feet high in a 200-foot-high tree she called "Luna," targeted by the Pacific Lumber Company as part of a large clearcut. The company tried to starve her out, blow her off with a helicopter's downdraft, and then deliberately cut trees so they would fall against hers. Ultimately an agreement was reached to spare Luna and Butterfly came down.

Scott Silver established Wild Wilderness in 1991 in the belief that Americans should not be required to "pay to play" on public properties they own and maintain through taxes. He found the issue goes deeper than admission charges to questions of proper use, and improper misuse, of public lands. In due course he became distressed over the congressionally authorized Recreation Fee Demonstration Program, strongly supported by commercial interests, especially snowmobile and other motorized off-road vehicle manufacturers and users, and over the failure of the "Big Greens" to oppose it.

I share Silver's concern. Yes, outdoor recreation spans a variety of interests, tastes, and goals. Commercial resorts and campgrounds bring the conveniences of urban living into outdoor life away from home, which is fine for those who want it that way. Disneyland and other profit-making theme parks furnish mass entertainment like television and movie theaters do. But public parks are like art galleries, museums, and libraries, meant to enrich society by enlightening and elevating individuals who come to them.

There is no way to place a dollar value on a "park experience" or a "wilderness experience," and yet the simple act of visiting the natural world has become a commercial transaction. Worst of all, the agencies in charge, the National Park Service and Forest Service, make "partnerships" with profit-driven entrepreneurs bent on introducing motorized forms of recreation and commercializing wilderness.

An Associated Press dispatch on August 7, 1999, cites the regional forester in the Southwest, Eleanor Towns, declaring: "Free enterprise in this region is alive and well." She was referring to approval of the transfer of 272 acres of the Kaibab National Forest at the gateway of Grand Canyon National Park in order to construct 1270 hotel rooms and 270,000 square feet of retail shopping, the equivalent of four large department stores. For his part, the park superintendent, Robert Arnberger, was all for creating the new city of Canyon Village as the companion piece to unlimited tourism at the Grand Canyon. Luckily, Navajo tribal members and other Arizonans objected with vehemence and blocked the project.

It's almost as though administrators get extra points for attracting crowds and commerce. Adolph Murie worked for the National Park Service for thirty-two years as a scientist, principally in Denali and Grand Teton National Parks. On November 8, 1956, he sent a memorandum to the park superintendent of McKinley (later renamed Denali) National Park, expressing concern over construction proposed in Mission 66, the ten-year development plan for the national park system. The park superintendent at the time, Duane D. Jacobs, brushed him off, saying: "It is quite reasonable for anyone of your many years of intimate knowledge of McKinley as purely a wilderness area to be somewhat alarmed as Mt. McKinley finally emerges across the threshold of a new era, that of a great national park set aside for the use and enjoyment of the people, which is soon to receive the intended use and enjoyment."

In 1973, Superintendent Jack Anderson of Yellowstone National Park received the First International Award of Merit from the International Snowmobile Manufacturers Association for his "enlightened leadership and sincere dedication to the improvement and advancement of snow-mobiling in the United States." Anderson certainly earned his award when he said that elk, bison, moose, even the fawns were unfazed by snowmobiles—in winter, the very season when they are the weakest and need to be left alone.

Agency officials claim fees are necessary to raise funds to protect natural resources. Then they place the burden on local administrators to serve as fee collectors and marketers of recreation as a commodity—that is how their performance will be rated. It's a terrible idea, thoroughly incompatible with principles of conservation.

I think it more fundamental to recognize that public lands serve as the basic reservoir for meeting outdoor recreation demands in America by persons of all economic levels. Even if the government takes in less from recreation sites than it costs to maintain them, recreation is not a net loss. To the contrary, the government's role in recreation should be to support conservation, physical fitness, and healthy outdoor leisure.

Scott Silver sparked substantial protest against recreation fees, so that extension of the fee demonstration program barely made it through Congress in 2001. Unfortunately, organizations like the National Parks and Conservation Association and the Wilderness Society were out to lunch—lunching with the wrong people.

Keeping the Faith and Friends

This little essay was not presented as a speech, but it seems fitting to assert the value of keeping faith and friends. Maybe that's the best lesson I've learned on my life journey, and it didn't come easily.

I arrived early, ahead of schedule, to help celebrate the first Earth Day at Earlham College in 1970. With time to spare in the lounge at the Quaker-supported college in eastern Indiana, I picked up a pacifist periodical, *Fellowship,* published by the Fellowship of Reconciliation. I was struck then and in years thereafter by the extensive treatment of environmental issues in this modest magazine and by the pacifist approach to social issues and to life.

Comfort comes from reassurance of the transcendent importance of each human being, no matter how obscure or difficult a life he or she may be living, Alfred Hassler wrote in the issue of *Fellowship* that I read at Earlham. He reminded me that God marks the fall of a single sparrow and grieves the suffering of the humblest of his or her children. But, Mr. Hassler continued, religion that only comforts is incomplete, for our paradise is fashioned on Earth, the here-and-now, the one part of the infinite in which man and woman can demonstrate themselves as worthy.

That makes sense to me. Critics like to put down environmentalists for treating their creed as a religion. But yes, it *is* a religion, founded on a belief that the earth and all its creatures are children of the same God, an appreciation of the mysteries of common origin and common destiny that bind us all together, and the challenge to care for each other. After all, "What should it profit a man if he gains the whole world and loses his own soul?"

Through *Fellowship* I learned Martin Luther King's Six Principles of Nonviolence, and they make sense to me too. Nonviolence is a way of life for courageous people. It seeks to win friendship and understanding. It seeks to defeat injustice, not people. It holds that suffering can educate and transform. It chooses love instead of hate. It believes the universe is on the side of justice and has faith that justice ultimately will prevail.

In *Fellowship* I read memorable essays by Sister Mary Evelyn Jergen, a prominent crusader for peace and social justice, and an interview with Sister Helen Prejean, author of *Dead Man Walking,* on her tireless campaign to end the death penalty. Sister Mary Evelyn cited forgiveness at the heart of nonviolence—requiring searching for the truth in the heart of the opponent, no matter how hidden from both parties. She wrote that we are all capable of more goodness and more evil than we can imagine: "This mysterious potential is frightening, and we tend to deny it. To withhold forgiveness, to act out of vengeance, reveals a refusal to accept the ignorance—our own and others'—that is part and parcel of life."

That is a hard lesson to learn. I confess it's been difficult for me in confrontations over environmental issues. I found help and encouragement in a different context early in 2001 when Morris Dees, an authentic American hero, came to speak in Bellingham. Dees, cofounder and director of the Southern Poverty Law Center, is undoubtedly the country's preeminent tracker of hate crimes, the American Elie Wiesenthal, renowned as the lawyer who successfully sued the Ku Klux Klan and the neo-Nazi Aryan Nations.

In his presentation Dees was anything but flamboyant or angry. He came across as serious and soft spoken. That was a lesson in itself: that I don't have to shout to be heard. "An ill wind is blowing across the nation," he said, citing 450 websites peddling hate via the Internet. Nevertheless, he insisted the principal mission of the Southern Poverty Law Center is not in the pursuit of legal action but in "teaching tolerance." He said that many racists and skinheads are homeless, powerless, and poor, people who need to be listened to, with friendship, understanding, and love, and with assurance that "we feel your pain." The title of his talk, in fact, was "Voices of Tolerance and Hope for the New Millennium."

There must be faith in his persona, and religion too, and the same for Sister Helen Prejean. When asked in her interview with *Fellowship* whether she had ever suffered from burnout, she replied simply that she had been too busy over the years on her mission, which she considered an assignment from God.

I won't presume any such mandate, but I have never felt burnout, the need to walk away and go do something else, not even through all the disappointments, defeats, and disillusionments. I'm not alone in feeling this way. I remember in the 1960s and 1970s when friends in California were campaigning to save the last remaining stands of redwoods, the tallest trees on Earth. They were ever so close to victory, yet the most devastating logging continued apace and big trees fell. A small national park was established in 1968, but ten years later, when Congress finally expanded the park with meaningful protection, the best of it was cut over and gone, leaving behind two hundred miles of logging roads, three thousand miles of skid trails, and thousands upon thousands burnt stumps of what were once magnificent redwood trees. It was heartbreaking, yet it led to a major rehabilitation program. Maybe one day redwoods will grow here again, but only because those who cared kept the faith through the darkest days.

Speaking of redwoods reminds me that in *Fellowship* I read about Julia Butterfly Hill, whom I mentioned earlier. She had prayed for some way to make a difference in the world, then for help to overcome the hatred she felt over destruction of great old trees. Prayer helped her survive enormous drenching and freezing storms. The worst storm almost killed her. The biggest gust threw her almost three feet. She grabbed the branch of the tree in the middle of her platform and prayed. She heard the voice of Luna tell her that trees in a storm let themselves go, allow themselves to bend and be blown with the wind.

When asking myself about avoiding burnout, I think of glorious places and wonderful people I've been lucky enough to know. For example, in the fall of 1975 I went to East Tennessee to help dedicate the Gee Creek Wilderness, which covers less than twenty-five hundred acres but presents a beautiful fragment of Appalachian mountain forest. Walter Williams, a Chattanooga graphic artist and conservationist, had discovered Gee Creek in his explorations of Cherokee National Forest, then worked with the local congressman, Lamar Baker, a conservative Republican, to lawfully review and establish the wilderness.

At the dedication, a commemorative marker was unveiled, with the date, October 4, 1975, and words from John Muir: "Climb the mountains and get their good tidings. Nature's peace will flow into you as sunshine flows into trees. The winds will blow their freshness into you and the storms their energy, while cares will drop off like autumn leaves." I think it's hard to read such lines in a setting to match and not be moved in an emotional and religious way.

Maybe another reason I avoided burnout is that I corresponded over the years with a lot of different people, sharing and learning in the process. Elsewhere in these pages I mentioned that Horace Albright made a practice of clipping and sending my column in the *Los Angeles Times* every Sunday with a note of comment or recollection. I was thrilled to have them and offered to pay the postage, but he declined, saying that would only spoil his pleasure.

I also treasure correspondence, and friendship, with John B. Oakes, editorial page editor of the *New York Times* for fifteen years until his retirement in 1976. He was a valedictorian at Princeton University and a Rhodes scholar, and awesome in his chosen profession. During the years he ran the editorial page Oakes editorialized about civil rights, the presidency, foreign affairs, politics, and the environment, defining a lofty agenda of public policy. When he retired his colleagues paid editorial homage (on January 1, 1977): "He could not and would not prettify the scene. But he dignified it, with a conscientiousness and with standards that were unyielding and with a boundless confidence that if only sound values and solid information could be located in the confusion of events, the citizen reader would distinguish the right from the wrong and uphold the public good."

Even after retiring, he contributed powerful, hard-hitting pieces about the environment to the op-ed page (which he had started in 1970). Personally, however, he was gentle, modest, and soft spoken. We had some discourse about objectivity. He wanted facts to speak for themselves, separating news from editorial and opinion. "It's a difficult distinction," he conceded, "but one cannot write an environmental story in the news columns without expressing a basic point of view of sympathy with the environmental viewpoint. It is impossible to report what's really going on and exposing what's going on without pointing out the anti-environmental actions being taken."

Following his death in April 2001 at the age of eighty-seven, I wrote a note of sympathy to his wife, Margery, and received a lovely reply: "Thank you for your letter, so appreciative of John. I hope you know how fully he reciprocated your feelings. It truly helps to hear from someone he valued as highly as he did you." Yes, that was nice to read, but it also made me acutely aware of the loss of a friend and hero.

In 1988 Oakes wrote me: "In any listing or any account of journalistic luminaries in the conservation field during the 50s, 60s or 70s, Dilliard [Irving Dilliard of the *St. Louis Post-Dispatch*] deserves a prominent place." Subsequently I corresponded with Dilliard, then living in Collinsville, Illinois, after retiring from the newspaper and from ten years as a professor at Princeton University in New Jersey. He wanted to tell me about Irving Brant (whom I have mentioned elsewhere) and Brant's many interests:

> That characterizes editorial writers. I had a thousand interests—still have—and the environment was one. Just as concerned about saving my land in the far west or Alaska as in the Ozarks.
>
> The last time I was with I.B. [Irving Brant] was Sunday, September 11, 1955, when he dedicated a bronze plaque at our Madison County Courthouse ten miles north of here in Illinois. He made the address of the day and I accepted for our citizens.
>
> Do you know the name of E. Palmer Hoyt? He was a Northwest editor whose career climaxed as publisher and editor of the *Denver Post,* beginning in 1946. Before that he had just about every advancing position on the *Portland Oregonian.* That, I'm rather sure, is where he and Dick Neuberger, one of my close friends, became acquainted. Incidentally, when Dick was invited to a major spot in New York, he asked me if he should take it. I told him to stay in the Northwest and spend even more time in Alaska and Canada. Hoyt *was an environmentalist!*

Through such correspondence I kept alive, involved and learning. In the course of my career I've made many friends, at different stations in life, and have tried to keep in touch with them. I made a speech once on this subject to a group of young environmental professionals in Washington, D.C., all eager and anxious to get ahead with their careers. I told them that when all is said and done, all you have left is your friends, so cultivate friendships and be true to them.

In 1970, in a very small way I helped Tom Bell when he was the first executive director of the Wyoming Outdoor Council and started the periodical *High Country News.* He has long been out of both those activities but we have remained friends. Early in 2001 he wrote to me from his home in Lander, Wyoming, that he liked my little saying, "Be of good cheer and all will be well," continuing:

> I think that Attila and the Huns are at the gate [reference to George W. Bush, Dick Cheney, and company]. If there is anything left in four years, we will pick up the pieces and go on. God help us. . . .
>
> Mike, I wish I had some of your good wisdom thirty years ago. But then again maybe it is just as well I didn't for I might have decided that big hill was too high to climb. It is amazing what you can do when you are young and foolish.

Maybe so, but Tom Bell is one of the heroes I've been privileged to know as a friend.

I will conclude my little sermon on faith and friendship by citing a piece of correspondence which is neither from nor to me. It was written by Brock Evans to Stewart Brandborg (whom I have mentioned earlier) in November 2000, but a copy was sent to me because we have all been friends and comrades working for the same cause for so long. Brock Evans grew up in Columbus, Ohio, and went to Seattle as a young lawyer, only to become Northwest field representative of the Sierra Club and in time vice president of the National Audubon Society and executive director of the Endangered Species Coalition. We met many times in many places. He was writing to Brandborg to congratulate him on receiving the Robert Marshall Award from the Wilderness Society, a vindication long overdue:

> I haven't seen you in many years, to my regret, although our paths have come close from time to time. But everywhere I go across the northern Rockies I hear that magic name, "Brandy," and hear that the fighting strong words have never wavered, and that the mighty heart has stayed centered and strong. . . .
>
> I will never forget how you, Brandy, picked up Rachel and me and our two infant sons at Dulles Airport

that cold January day in 1973, when we had just moved to Washington to head up the Sierra Club office. Cold, tired, scared, and sad about leaving our safe secure Seattle home and friends, you just took us under your wing and with that wonderful irrepressible smile made me think it would be all right. . . .

So, congrats, again, and many, many thanks not only for all I learned from you, but for being the kind of role model for so many of us that you still are.

I Must Be Still to Hear and See

At the Northwest Wilderness and Parks Conference,
Seattle, October 8, 1994.

Everything is green. Everything is waiting and still.
Slowly things begin to move, to slip into their place.
Groups and masses and lines tie themselves together.
Colours you had not noticed come out, timidly or
boldly. In and out, in and out your eye passes. Nothing
is crowded; there is living space for all. Air moves
between each leaf. Sunlight plays and dances. Nothing
is still now. Life is sweeping through the spaces. Everything is alive. The air is alive. The silence is full of
sound. The green is full of colour. Light and dark chase
each other. Here is a picture, a complete thought, and
there another and there. . . . There are themes everywhere, something sublime, something ridiculous, or
joyous, or calm, or mysterious. Tender youthfulness
laughing at gnarled oldness. Moss and ferns, and leaves
and twigs, light and air, depth and colour chattering. . . .
You must be still in order to hear and see.

I am stilled within by contemplating those words, a word picture,
from Emily Carr's journal, written in the course of sketching and painting her powerful impressionistic portraits and landscapes of totem
poles, and of trees, tree trunks and wild forests of British Columbia.
Emily believed the glories she found and felt were derived from spiritual
sources. But then creative persons forever seek the primeval as source
material and inspiration. Literature, poetry and science, as Emerson
wrote, all are homage to the unfathomed secrets of nature. Ansel Adams
explained it this way: "Here are worlds of experience beyond the world

of aggressive man, beyond history, beyond science. The moods and qualities of nature and the relations of great art are difficult to define; we can grasp them only in the depths of our perceptive spirit."

This type of experience is still available for me, but I recognize that I must be still in order to hear and see. I must go to the sanctuary in the wild, willing to meet it on its terms, rather than on my terms or those of my own mechanistic society. Wilderness is the sacred place for renewal and healing, where "education" and "recreation" take on different meaning. A psychologist might prescribe a wilderness experience because of its freedom from evidences of critical or harmful human actions, or to find release from stress through stillness and solitude in the primeval. There are no social values to conform to; it is classless—all parties become essentially equal, benefiting from cooperation rather than competition. The individual acquires a sense of scale, conceding there is something larger and longer-lasting than anything he or she has known before and feeling that he or she belongs at the bosom of a much greater whole—and at peace.

The environmental model provides a measure of restorative impetus individuals need under current circumstances of alienation from nature and from each other. Shakespeare summed it up in *As You Like It,* when the banished Duke Senior struck a contrast between virtuous country life and the decadence of the court. "And this our life, exempt from public haunt,/finds tongues in trees, books in the running brooks,/sermons in stones and good in everything," the Duke said (in act 2, scene 1), clearly finding nature's reality a singular comfort and spiritual guide.

"The one thing in the world, of value, is the active soul," wrote Emerson. "This every man [and woman] is entitled to; this every man [and woman] contains within him [and her], although, in almost all men [and women], obstructed, and as yet unborn." As in the ancient sacred places of all religions, as in the sacred sites that hold meaning still to native Americans, wilderness evokes the active soul, freeing it to respond to the earth as alive, poetic, dramatic, musical. Wilderness breaks down artificial barriers between people bred to believe they are different from each other by reason of class, color, race, or gender; wilderness is teaching, real teaching, through which physically and mentally disabled learn to overcome limitations, and the abled learn to think differently about competency. But, of course, in a sacred place all life is sacred, and the humblest are holy and blessed.

National parks are not playgrounds, nor theme parks, but sanctuaries, meant to be forever; they are priceless time capsules for tomorrow that we are privileged to know and enjoy today. By that I mean a national park is ideally suited to exercise the body in a test with nature, stimulate the mind with new learning, and challenge the spirit, the spirit of the individual to connect with something larger than himself or herself, and more lasting than all the mechanization of life and work at home.

As evidence, I cite the experience of Mark Wellman and the lessons from it. Mark was an accomplished California mountaineer who broke his back in a climbing accident in 1982 and was left without the use of legs. He lost direction in his life, lived in pain, loneliness, and shattered dreams—until he found his new beginning in Yosemite. Living and working in the park, Mark pushed himself to see as much as humanly possible in a wheelchair. He advanced bit by bit, building his upper body, ultimately making history in 1991 when he and a partner climbed the thirty-five-hundred-foot granite face of El Capitan. Then two years later he pulled himself to the summit of Half Dome, though it took thirteen days to make it. "I've always believed that true adventure involves discovering things about yourself as you edge ever closer toward the boundaries of your personal limits," Mark wrote later. "I learned plenty about myself on El Capitan and Half Dome."

National parks *are* true adventure, places for discovering things about oneself, for edging toward boundaries of personal limits. It doesn't have to be intensely physical either. Walk only a short distance from the main paths. In solitude away from crowds I encounter and examine flowers, trees, birds, rocks, and water—separately and then collectively. It's astonishing how I can train my eyes, ears, and nose to note things others ignore. Looking at scenery can be a passive experience, but as I explore with the eye and mind, patterns of nature become evident and logical. I can become my own ecologist and enriched in spirit in the process. That is using the parks as they were meant to be used.

Doing so is personally rewarding and benefits the parks cause. We Americans love to travel, when, where, and however we want. I hate to moralize, or to advocate strict rules and regulations, or restraints on individual freedom in the out-of-doors, but with this freedom of mobility comes the responsibility to protect the environment and the ability of others to travel freely. In return for what they give, national parks and other public lands deserve and need the respect of visitors who come to enjoy them. I believe that Americans make mistakes in the out-of-doors without malice. Wilderness is our common treasure and trust. When the problems derived from overuse and abuse are explained properly, Americans will understand and respond appropriately, and, hopefully, influence the body politic that serves us.

Overuse and misuse clearly deplete the visible physical resource that people care about, but it does something to the invisible spirit of place as well. Native Americans have that ancestral sense, honoring the earth and life as divine gifts. Here on the Northwest Coast native people for centuries have sought the giant cedar, hemlock, and Douglas fir of the cold rain forest, not simply for canoes and longhouses, but as source of "a sacred state of mind where magic and beauty are everywhere." In *The Vanishing American,* Zane Grey's hero, Nophaie, most loved to be alone, out in the desert, "listening to the real sounds of the open and to the whispering of his soul." Grey wrote that Nature was jealous of her secrets and spoke only to those who loved her. The Rainbow Bridge, just north of the Arizona–Utah border, curving upward to a height above three hundred feet, once was a sacred destination for religious pilgrimage, reached by toil, sweat, endurance, and pain, proving to the pilgrim that the great things in life must be earned. That makes sense even to the European mind, for as Jung wrote, "There is no birth of consciousness without pain." Now, by contrast, however, the impounded waters of Glen Canyon Dam have made painless visits possible, via boat on the reservoir called Lake Powell to the Bridge Canyon landing, then walking about one mile. Surely some element of critical value—the sense of connecting with spirit—is lost.

Sacred places are located widely across North America, retreats where native peoples in many different ways have sought to cleanse body, mind, and spirit, to experience visions, revelations, mystic journeys. I can hardly resist in passing to speak in support of the Native American Cultural Protection and Free Exercise of Religion Act, now pending in Congress, to safeguard these sites on federal lands. The whole country will benefit from it.

I recognize that sacredness sustained in wild places cannot provide a quick fix for the ills of society, but it can bring new understanding at a very personal level. Society divides by economics, race and religion, gender and sexual preference, and by physical ability. Much of the time people feel separate and fearful, as children of different gods, of greater and lesser gods. Wilderness evokes the unity and wholeness of creation, a community of brothers and sisters all kin, all equal, sharing in common origin, common destiny.

I learned this anew in 1992 in the Boundary Waters of Minnesota, traveling with Wilderness Inquiry, a nonprofit group that mixes people with and without disabilities; I mean that participants include wheelchair users, the deaf and blind, and the mentally retarded, and that it works.

Wilderness on our trip became more than a physical or intellectual experience, but also a place to think differently about physical competency and accomplishment. We worked on potentials rather than limitations, while paddling, portaging, doing camp chores, sharing songs and sunsets. Committed climbers and kayakers may prefer wilderness where they expose themselves to physical risk—that may be their sacred space—but when people of mixed ability live, work, and play together, they expose themselves, too, learning to rise above the barriers of modern society, and to recognize that every life has meaning to it.

Disabled are subject to discrimination everywhere, reminded of their limitations, of how they are different, but the mixed-ability wilderness trips show that everyone has limitations, and that every person has something to give that another needs or wants. On most outfitter-led trips, individuals have things done for them; they are carried along on the strength of the leader, or of the group. I contrast the Wilderness Inquiry mode with the climbing concession at Mount Rainier National Park, operating like an assembly line, shepherding the flock to the summit for a mighty macho exultation. The outdoors has long been dominated by the male macho world view, but I found that actual physical accomplishment counts less than the effort, that an able-bodied person carrying a heavy pack may be worth less than a mobility-impaired person crawling across the trail.

Time spent in wilderness benefits anyone and everyone, that is true, but everyone doesn't have to come at the same time. Thirty years ago Dr. Bettie Willard recorded the effects of intensive human use on the alpine vegetation of Rocky Mountain National Park. She found areas where, after decades of trampling in the short summer, all plants and five inches of topsoil were gone, leaving a bare mineral surface. Willard estimated that a minimum five hundred years would be necessary to restore the tundra ecosystem.

In the history of the earth, five hundred years is not so very long. In the meantime, we need to walk lightly, *more* lightly, and, hopefully, to learn that the gods walk on every road and every road is sacred.

One More for the Road

I thought that somehow, considering my age and outspokenness, this might be the last speech I ever gave, so I wanted it to be memorable, at least to me, and to my wife, my love-mate and partner in the search for a better world. I was home again in Idaho, the most politically conservative of all the states, but I felt free of restraints on what I could say or what I could not say.

My old colleagues had invited me back to lecture in classes and to deliver a speech at what they called a "campus/community assembly." They had reached out to involve as sponsors various departments on campus, plus citizen groups in town, including Friends of the Clearwater, the Idaho Conservation League, the local Sierra Club chapter, Palouse Clearwater Environmental Institute, and the Earth and Faith Forum. It took some courage, for that kind of liaison in higher education is rare.

But it filled the hall, with a crowd of four hundred overflowing the University of Idaho Law Auditorium. My colleagues had asked me to provide a title for my talk, which became "Who Owns the Earth? Who in the End Will Save It?" And now, on October 23, 2001, the title became particularly fitting.

The nation and world were still in shock over the bombing of the World Trade Center in New York only six weeks before. I thought the wounds were still too raw for me to say that the United States itself had spread terrorism around the world, and that the politics of globalization had led our country to support corrupt, cruel, and dictatorial regimes. But I sought to get the message across by giving my vision of the world the way I want to see it.

I said that my vision begins in a world where children and women are safe; where riches are distributed with equity—the riches of spirit as well as of substance, the riches of giving and caring and sharing—in a world where all life is venerated; where resources are no longer wasted on bombs and wars but are directed to the useful and enriching, such as schools, libraries, playgrounds, parks, health centers, theaters, concert halls, and art galleries; where we the people are empowered as decision makers in an open, responsive system; and where wild places are recognized as sacred places for pilgrimage and personal transformation.

In the twenty-first century we live on a planet deeply wounded, troubled by terrorism and its root causes, particularly poverty on a very, very large scale. The world, after all, is divided between those who do not have enough, including many who are hungry and hopeless, and those who have more than enough and still demand more. And the whole world, the rich and poor of it, is plagued by global warming, acid rain, destruction of tropical forests, loss of wildlife, toxic wastes, poisoned air and water, and the pressures of growing population.

In my worldview the question of "Who owns the earth?" is best answered by citing the principle of *usufruct*. In the United States the idea of private property rights generally prevails. The earth has a price tag on it; private profit transcends public responsibility. But usufruct is a much older and better idea, based on stewardship. Usufruct is a philosophical/legal ideal born in ancient times, when land, crops, gardens, and livestock were often used communally. In so-called primitive societies, land and its resources were merely borrowed from future generations and must be passed on intact—a social pact of stewardship. Thomas Jefferson strongly advocated this principle, writing: "Each generation has the usufruct of the earth during the period of its continuance. When it ceases to exist, the usufruct passes on to the succeeding generation free and unencumbered and so on successively from one generation to another, forever."

I felt free in my speech to say that I have even chosen theme music to accompany my world view, or vision, namely Gustav Mahler's Symphony Number Three in D Minor. It is performed by a full orchestra and chorus of 150 voices, and I value it for its statements in music of a philosophy of humankind and its relationship to nature, of the relationship between art and life, and even of the social equality of all people. As Mahler described the first movement (which he composed last), it is a "gigantic hymn to the glory of every aspect of creation . . . and to the miracle of spring, thanks to which all things live, breathe, flower, sing, and ripen, after which appear those imperfect beings who have participated in this miracle—the men [and women] . . . right up to the kingdom of the spirit and that of the angels."

And I find Mahler's aesthetic wholly compatible with Gifford Pinchot's assertion that "the earth belongs by right to all its people and not to a minority, insignificant in number but tremendous in wealth and power."

Pinchot learned the hard way over time. In his autobiography, *Breaking New Ground*, he wrote about going west late in the last years of the nineteenth century, learning about the Western forests, and about the big men and little men who used and abused them. The powers and principalities which controlled the politics and the people of the West emerged for him from the general landscape:

> Principalities like the Homestake Mine in the Black
> Hills, the Anaconda Mine in the Rockies, Marcus Daly's
> feudal overlordship of the Bitterroot Valley, and Miller
> and Lux's vast holdings of flocks and herds and control
> of grazing lands on the Pacific slope—these and others
> showed their hands or their teeth. So did powers like
> the Northern and Southern Pacific and the Great North-
> ern Railroads, the irrigation interests of California, and
> the great cattle and sheep stock growers' associations.

In a sense, little has changed. I told my Idaho audience that in our own time corporate influence has grown more blatant. Indeed, the George W. Bush administration is probably the most openly anti-environmental administration in a century, with industry people appointed to key positions throughout the government and serving the corporate cause.

Consider that the United States, with 4 percent of the world's population, is responsible for 25 percent of the world's carbon dioxide emissions. Yet George W. Bush rejected the global warming treaty, claiming we don't have enough scientific evidence and the treaty is unfair to the U.S. He took $200 million from renewable energy and energy efficiency programs. He cut by 50 percent funds for research into wind, solar, and geothermal energy. This is not hard to explain, for the head of his federal energy policy task force earned $30 million a year as an oil company executive—his vice-president, Dick Cheney.

No good will come of it. Such corporate influence weakens and paralyzes government and professional agencies at all levels. Federal lands are in dismal condition today—timberlands are overcut, rangelands overgrazed, fisheries depleted, rivers overdammed, parks abused and overused, subsidized mining degrades landscapes.

We should not allow it, for public lands are the heart and body of the West, and maybe soul too. Public lands are the source of Western art, literature, history, and mystery, the mystery of people trying to relate to the earth around them, failing at times, and falling, picking up to try again. Take away the public lands from the environs of Albuquerque, Boise, Denver, Salt Lake City, and Seattle, and they would be the most ordinary of places. Take away the public lands and there wouldn't be much to the economy either. Public lands are the last open spaces, last wilderness, last wildlife haven. Without public lands the West would be an impoverished province.

The most important and legitimate role of government is to protect and care for its citizens, especially the weak, sick, and vulnerable. Good government insures that every citizen has access to housing, food, health care, and education. That is exactly what Pinchot preached and worked for. But politicians of both parties have pursued an agenda to consolidate wealth and power, power and wealth, with government the agent of the powerful few. Or, as Supreme Court Justice Louis Brandeis wrote in the 1920s, "We can have democracy in this country, or we can have great wealth concentrated in the hands of a few, but we cannot have both."

Still, I believe I can change the world. I can do this by changing my own image of reality, by getting rid of poisonous beliefs that have led to the current state of affairs. I can do it by joining with others who foresee the transformation of society. The future will be different if we make the present different. The thing to do right now is to create a new society within the shell of the old with the philosophy of the new, which is not a new philosophy but a very old philosophy, a philosophy so old that it looks new.

Wilderness is at the core of the healthy society. Wilderness, above all its definitions, purposes, and uses, is sacred space, with sacred power, the heart of a moral world, a way of understanding the sacred connection with all of life.

All across the world so-called primitive peoples have placed a personalized value on the sacred qualities of land, particularly when untouched. Their sacred places serve as a common history uniting generations, where they cleanse body, mind, and spirit, experience visions, revelations, mystic journeys, praying and fasting, ritually washing in sacred streams.

Native Americans have that ancestral sense, honoring the earth and life as divine gifts, even after four centuries of genocide, and all the accompanying discrimination, degradation, and despair. The past for them lives in the present: Land, water, trees, animals, birds, trees, rocks, and human remains are believed instilled with vital and sacred qualities. On the Northwest Coast where I live native, people for centuries have sought the giant cedar, hemlock, and Douglas-fir of the cold rain forest, not simply for canoes and longhouses, but as source of a sacred state of mind where magic and beauty are everywhere.

They want to be part of a modern world, while kindling and rekindling earth-based tradition. They live in the belief that better days will come, if they can mend the sacred hoop broken many times in the past centuries. For the Chippewa, for example, the wolf is brother and friend, the teacher of honor, endurance, perseverance, and loyalty. Their belief and tradition say that the creator placed the wolf to walk with the first man of the Chippewa. Then the creator told them that what happens to one will happen to the other. Now, in our time, Chippewa leaders say: "This has come to pass. We've had our lands taken, been hunted for our hair, and pushed to extinction. But we see the wolf returning, gaining strength in places where he was once destroyed. Perhaps the wolf will lead the way to more natural living and teach respect for Mother Earth."

Crusades for social issues, whether for peace, racial equality, gender rights, or the environment, show how people—at times a very few—can and do bring needed change. Moreover, the effort itself is rewarding, the change in oneself matters most, more than whatever success the effort may bring. I. F. Stone, the hero of courageous and fearless journalism, said it well in December 1971 at the close of his career:

> The place to be is where the odds are against you; power breeds injustice, and to defend the underdog against the triumphant is more exhilarating than to curry favor and move safely with the mob. Philosophically I believe a man's life reduces itself ultimately to a faith—the fundamental is beyond proof—and that faith is a matter of aesthetics, a sense of beauty and harmony. I think every man is his own Pygmalion, and

spends his life fashioning himself. And in fashioning himself, for good or evil, he fashions the human race and its future.

Rachel Carson, in reviewing her life and goal, felt much the same: "The beauty of the living world I was trying to save has always been uppermost in my mind—that, and anger at the senseless brutish things that were being done. I have felt bound by a solemn obligation to do what I could—if I didn't at least try I could never again be happy in nature."

Ernie Dickerman, who fought for wilderness and the out-of-doors, was never discouraged. "It is amazing," he wrote in 1972, "how political democracy in the United States, despite its deficiencies and innumerable errors, permits so many of us to lead satisfying, rewarding lives."

For my last citation I turn to the all-American children's hero, Mr. Rogers. When he retired from public television in 2001, Mr. Rogers said: "At the center of the universe is a loving heart that continues to beat and that wants the best for every person. Anything we can do to help foster the intellect and spirit and emotional growth of our fellow human beings, that is our job. Those of us who have this particular vision must continue against all odds. Life is for service."

As for myself, I see imagination and a subjective value system as a force empowering the individual who truly cares. To say it another way, the source of strength in human life is in emotion, reverence, and passion, for the earth and its web of life. Life *is* for service, and each individual must realize the power of his or her own life and never sell it short.

Index

Greenspeak was designed and typeset on a Macintosh computer system using QuarkXPress software. The text is set in Garamond ITC, and the chapter openings are set in Helvetica Neue. This book was designed by Cheryl Carrington, typeset by Kimberly Scarbrough, and manufactured by Thomson-Shore, Inc.